Exam Preparation
Electrotechnical apprenticeship qualification (5357)

Exam Preparation
Electrotechnical apprenticeship qualification (5357)

Published by The Institution of Engineering and Technology, London, United Kingdom

The Institution of Engineering and Technology is registered as a Charity in England & Wales (no. 211014) and Scotland (no. SCO38698).

This publication is copyright under the Berne Convention and the Universal Copyright Convention. All rights reserved. Apart from any fair dealing for the purposes of research or private study, or criticism or review, as permitted under the Copyright, Designs and Patents Act 1988, this publication may be reproduced, stored or transmitted, in any form or by any means, only with the prior permission in writing of the publishers, or in the case of reprographic reproduction in accordance with the terms of licences issued by the Copyright Licensing Agency. Enquiries concerning reproduction outside those terms should be sent to the publishers at The Institution of Engineering and Technology, Michael Faraday House, Six Hills Way, Stevenage, SG1 2AY, United Kingdom.

Copies of this publication may be obtained from:
The Institution of Engineering and Technology,
PO Box 96,
Stevenage,
SG1 2SD, UK
Tel: +44 (0)1438 767328
Email: sales@theiet.org www.electrical.theiet.org/books

While the author, publisher and contributors believe that the information and guidance given in this work are correct, all parties must rely upon their own skill and judgement when making use of them. The author, publisher and contributors do not assume any liability to anyone for any loss or damage caused by any error or omission in the work, whether such an error or omission is the result of negligence or any other cause. Where reference is made to legislation it is not to be considered as legal advice. Any and all such liability is disclaimed.

The moral rights of the authors to be identified as author of this work have been assertedby them in accordance with the Copyright, Designs and Patents Act 1988.

ISBN 978-1-78561-567-2 (paperback)
ISBN 978-1-78561-568-9 (electronic)

Typeset in India by S4Carlisle Publishing Services
Printed in the UK by Greens Ltd, Lincoln Road, Cressex Business Park, High Wycombe, Buckinghamshire HP12 3RQ

Contents

Introduction	8

The exam

Test specifications	10
The e-volve platform	17
Examiner's Tips: Preparing for the exam	19
Examiner's Tips: During the exam	19
Tips regarding the questions	20
Written exams	22
Pull out the information from the question or scenario	23

Exam Practice 1

Exam practice 1: Unit 101/001 understand health, safety and environmental considerations	26
Exam practice 1 questions with answers: Unit 001 understand health, safety and environmental considerations	32
Exam practice 1 answer key: Unit 001 understand health, safety and environmental considerations	41

Exam Practice 2

Exam practice 2: Unit 003 electrical scientific principles and technologies	44
Exam practice 2 questions with answers: Unit 003 electrical scientific principles and technology	56

Exam practice 2 answer key: Unit 003 electrical scientific principles and technologies	76

Exam Practice 3

Exam practice 3: Unit 103 (written paper) electrical scientific principles and technologies	78
Exam practice 3 questions with model answers: Unit 103 written paper	82

Exam Practice 4

Exam practice 4: Unit 004 understand design and installation practices and procedures	96
Exam practice 4 questions with model answers: Unit 004 understand design and installation practices and procedures	104
Exam practice 4 answer key: Unit 004 understand design and installation practices and procedures	118

Exam Practice 5

Exam practice 5: Unit 005 understand how to plan and oversee electrical work activities	120
Exam practice 5 questions with model answers: Unit 005 understand how to plan and oversee electrical work activities	124
Exam practice 5 answer key: Unit 005 understand how to plan and oversee electrical work activities	130

Exam Practice 6

Exam practice 6: Unit 012 understand inspection, testing and commissioning	132

EXAM PREPARATION
CONTENTS

Exam practice 6 questions with model answers: Unit 012 understand inspection, testing and commissioning — **140**

Exam practice 6 answer key: Unit 012 understand inspection, testing and commissioning — **152**

Exam Practice 7

Exam practice 7: Unit 112 written paper understand inspection, testing and commissioning — **154**

Exam practice 7 questions with model answers: Unit 112 written paper understand inspection, testing and commissioning — **158**

Exam Practice 8

Exam practice 8: Unit 014 understand fault diagnosis and rectification — **166**

Exam practice 8 questions with answers: Unit 014 understand fault diagnosis and rectification — **173**

Exam practice 8 answer key: Unit 014 understand fault diagnosis and rectification — **184**

More Information

Further reading — **186**

Online resources — **187**

Further qualifications — **188**

Notes

Introduction

How to use this book
This book has been written as an exam practice aid for some of the exams you will need to take to complete the City & Guilds Level 3 NVQ Diploma in Electrotechnical Technology (5357). It sets out methods of studying; offers advice on exam preparation and provides details of the scope and structure of the examinations. You will also find guidelines and advice about sitting the exam.

There are practice examinations in this book, and you will also find fully worked and 'model' answers. Used as a study guide for exam preparation and practice, it will help you to reinforce and test your existing knowledge.

Try to answer the sample test questions under exam conditions, and then review all of your answers. This will increase your familiarity with the different types of question that might be asked in the exam. It will also help you understand how long the questions take, and learn how to pace yourself in order to complete all questions in the time allowed for the exam.

This book cannot guarantee a positive exam result, but it can play an important role in your revision, helping you to prepare for and approach the exam confidently.

About City & Guilds Level 3 Electrotechnical Apprenticeship Qualification (5357)
The qualification you are taking is for learners who want to work as an electrician, installing systems and equipment in buildings, structures and environment. It allows you to acquire the necessary knowledge and practical skills regarding the design, installation and commissioning of electrical systems. Once you have demonstrated the required knowledge and site-based performances within this qualification, you will be eligible to take the AM2 assessment to become fully qualified.

With that qualification, you will also become eligible to apply for Engineering Technician status through the IET. This means that you will be professionally registered with the Engineering Council, and once registered you will be able to use the initials 'EngTech' after your name.

The exam

Test specifications	10
The e-volve platform	17
Examiner's Tips: Preparing for the exam	19
Examiner's Tips: During the exam	19
Tips regarding the questions	20
Written exams	22
Pull out the information from the question or scenario	23

Test specifications

This qualification is assessed using a range of assessments including practical assessments, written examinations and online multiple-choice examinations. Each examination, whether it be written or multiple-choice will have a test specification. These test specifications show the following information:

- Assessment method e.g. multiple-choice/written
- Examination duration
- Permitted materials e.g. closed/open book
- Number of questions
- Approximate grade boundaries

The grade boundaries may be subject to slight variation to ensure fairness should any variations in the difficulty of the test be identified.

The way the knowledge is covered by each test is laid out in the Test Specifications below:

001 Understand Health, Safety and Environmental Considerations

	Outcome	Number of questions	%
1	Understand how relevant legislation applies in the work place	4	16
2	Understand the procedures for dealing with Environmental and Health and Safety situations in the work environment	6	24
3	Understand the procedures for establishing a safe working environment	7	28
4	Understand the requirements for identifying and dealing with hazards in the work environment	8	32
	Total	**25**	**100**

Assessment method: e-volve online; multiple-choice test; closed book
Duration: 40 mins
Grade boundaries: Pass is approximately 60%

THE EXAM
TEST SPECIFICATIONS

003 Electrical Scientific Principles and Technologies

	Outcome	Number of questions	%
1	Understand mathematical principles which are appropriate to electrical installation, maintenance and design work	2	5
2	Understand standard units of measurement used in electrical installation, maintenance and design work	5	13
3	Understand basic mechanics and the relationship between force, work, energy and power	7	18
4	Understand the relationship between resistance, resistivity, voltage, current and power	15	37
5	Understand the fundamental principles which underpin the relationship between magnetism and electricity	7	17
6	Understand the types, applications and limitations of electronic components in electrotechnical systems and equipment	4	10
	Total	**40**	**100**

Assessment methods: e-volve online; multiple-choice test; closed book
Duration: 90 mins
Grade boundaries for this test will approximately be:
Pass: 50%
Merit: 65%
Distinction: 80%

Notes

103 Electrical Scientific Principles and Technologies

	Outcome	Number of questions	%
7	Understand electrical supply systems	6	23
8	Understand how different electrical properties can affect electrical circuits, systems and equipment	8	31
9	Understand the operating principles and applications of DC machines and AC motors	4	16
10	Understand the operating principles of electrical components	3	11
11	Understand the principles and applications of electrical lighting systems	3	11
12	Understand the principles and applications of electrical heating	2	8
	Total	**26**	**100**

Assessment methods: Short answer question paper
Duration: 120 mins
Grade boundaries for this test will approximately be:
Pass: 50%
Merit: 65%
Distinction: 80%

THE EXAM
TEST SPECIFICATIONS

004 Understand Design and Installation Practices and Procedures

	Outcome	Number of questions	%
1	Understand how to prepare for the installation of wiring systems	5	17
2	Understand the applications of wiring systems	9	30
3	Understand the practices and procedures for carrying out electrical work	8	26
6	Understand protection against overcurrent	5	17
7	Understand electrical systems and circuits	3	10
	Total	**30**	**100**

Assessment methods: e-volve online; multiple-choice test/written task; open book
Duration: 70 mins
Grade boundaries: Pass is approximately 70%

005 Understand how to plan and oversee electrical work activities

	Outcome	Number of questions	%
1	Understand the requirements for liaising with others when organising and overseeing work activities	10	63
3	Understand the requirements for organising the provision and storage of resources that are required for work activities	6	37
	Total	**16**	**100**

Assessment methods: e-volve online; multiple-choice test/written task; closed book
Duration: 40 mins
Grade boundaries: Pass is approximately 60%
Note: Learning outcome 2 will be assessed through an assignment

Notes

012 Understand Inspection, Testing and Commissioning

	Outcome	Number of questions	%
1	Understand the requirements for completing the safe isolation of electrical circuits and installations	6	17
2	Understand the requirements for initial verification of electrical installations	2	6
3	Understand the requirements for completing the inspection of electrical installations prior to their being placed into service	Assessed by assignment	
4	Understand the requirements for the safe testing and commissioning of electrical installations	8	23
5	Understand the requirements for testing before circuits are energised	7	20
6	Understand the requirements for testing energised installations	10	28
7	Understand the requirements for the completion of electrical installation certificates and associated documentation	2	6
	Total	**35**	**100**

Assessment methods: e-volve online; multiple-choice test; closed book
Duration: 120 mins
Grade boundaries: Pass is approximately 60%

THE EXAM
TEST SPECIFICATIONS

014 Understand Fault Diagnosis and Rectification

	Outcome	Number of questions	%
1	Understand the Health and Safety requirements relevant to fault diagnosis	3	10
2	Understand the importance of reporting and communication in fault diagnosis	2	7
3	Understand the nature and characteristics of electrical faults	6	20
4	Understand the fault diagnosis procedure	10	33
5	Understand the procedures and techniques for correcting electrical faults	9	30
	Total	**30**	**100**

Assessment methods: e-volve online; multiple-choice test; closed book
Duration: 120 mins
Grade boundaries: Pass is approximately 60%
Note: learning outcome 6 will be assessed through a practical observation

EXAM PREPARATION

Notes

016 Understand the Requirements of Electrical Installations BS 7671: June 2008 (2015)

	Outcome	Number of questions	%
1	Know the scope, object and fundamental principles of BS 7671	4	7
2	Understand the definitions used within BS 7671	2	3
3	Understand how to assess the general characteristics of electrical installations	6	10
4	Understand requirements of protection for safety for electrical installations	15	25
5	Understand the requirements for selection and erection of equipment for electrical installations	14	23
6	Understand the requirements of inspection and testing of electrical installations	4	7
7	Understand the requirements of special installations or locations as identified in BS 7671	10	17
8	Understand the information contained within the appendices of BS 7671	5	8
	Total	**60**	**100**

Assessment methods: e-volve online; multiple-choice test; open book
Duration: 120 mins
Grade boundaries: Pass is approximately 60%

Practice tests for this assessment are not contained within this book. Please refer to publications or practice tests for City & Guilds 2382.

The e-volve platform

All City & Guilds multiple-choice online tests are taken using the e-volve platform. In this section you will see some screenshots of e-volve to help you familiarise yourself with the e-volve tests.

Signing in

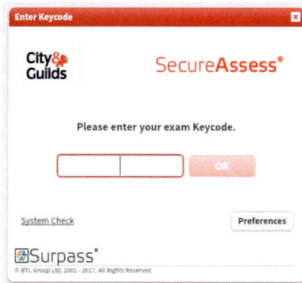

Once you have signed in using your keycode, you can change preferences

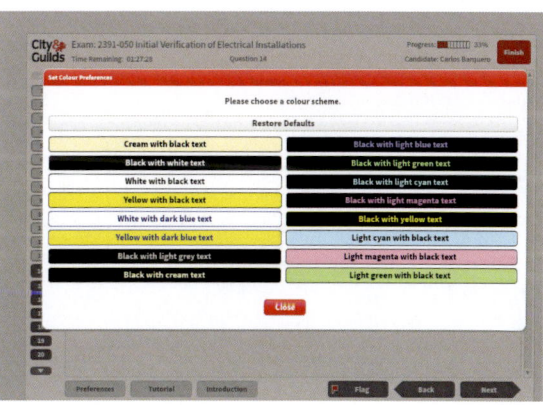

Some people prefer to read text using different colours or different backgrounds. For example, some people with dyslexia find reading black text on a yellow background much easier. If you do not set preferences, the default will be black text on a white background. You can change your preferences at any point during the examination.

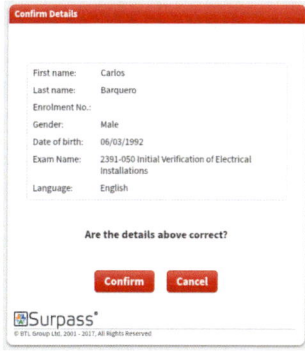

You will then be shown a screen with your details. Ensure they are correct as you wouldn't want to pass an exam for someone else!

Notes

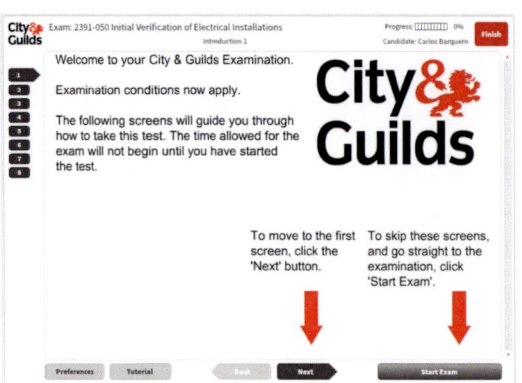

The welcome screen will give you options to either take a practice familiarisation session, or to carry on and start your examination. The clock doesn't start until you start the actual exam so if you are not familiar with the features of an e-volve test, use the *next* button.

When taking the examination, the screen has several features:

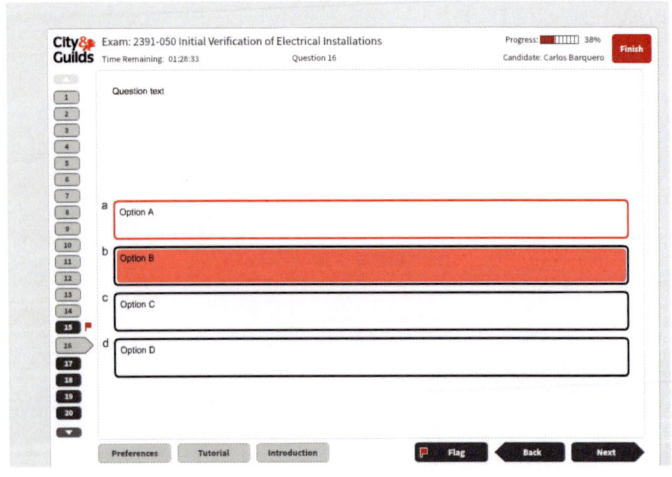

- **Timer** – this shows the time remaining for the examination.
- **Progress** – gives you an indication of how much of the examination you have completed.
- **Finish** – only click this button when you are sure you have completed the examination.
- **Question Numbers** – the numbers at the side show the individual question numbers. These indicate the questions that have been answered or the ones that aren't completed. They also indicate those questions that have been flagged (see below). You can navigate through the questions by clicking on each number or by using the *back* and *next* buttons at the bottom of the screen.
- **Flag** – if there are any questions you had doubts about, click the *flag* button and a flag will appear in the question number. This means you can easily navigate back to that question should you have time.
- **Preferences** – throughout the examination you can change the screen preferences.
- **Help** – unfortunately, this will not help you with the examination questions, only with the features of the system.
- **Introduction** – takes you back to the welcome screen.

THE EXAM
EXAMINER'S TIPS: PREPARING FOR THE EXAM

Examiner's Tips: Preparing for the exam

✓ Make sure you revise before the exam. Use practice exams, such as the ones in this book, to identify areas of weakness and focus your revision on these.

✓ Check the test specification to see what areas the exam focuses on. Check the duration and number of questions in the exam.

✓ Be sure you know what time the exam starts. The exam begins at that time; not at the time you arrive!

✓ Arrive at least 30 minutes before the start time so you can settle, check you have everything you need and ensure you calculator works.

✓ Make sure you go to the toilet. If you end up going during the exam, you will lose time and disrupt others.

✓ Make sure you know the requirements for the exam. Is the exam open book? What books can you take in? Do you need a calculator? Always take in pens or pencils and ask for scrap paper from the invigilator when you go in. It always helps to write things down if you need to perform calculations or take key information from a structured question.

Examiner's Tips: During the exam

✓ Listen carefully to instructions given to you by your invigilator.

✓ Attempt to answer every question. Leaving a multiple-choice question unanswered will be marked as incorrect. Have an educated guess as you have a 25% chance of being right.

✓ Read the question and, for multiple-choice, every answer before you attempt the question. If you read it too quickly, you may miss key words or information.

✓ Do not rush. You have, on average, two minutes per question.

✓ Do not worry about answering questions in order, you can leave a question and go back to it.

Notes

Notes

✓ Never let a question get to you! If a single question is not working out or you cannot find an answer, '**flag it**', have a guess and move on. If one question unsettles you, the remaining questions will become much harder because your state of mind has changed. Save the remaining questions by moving on **before** you unsettle. If you flag the question, you can easily go back if you have time at the end. Sometimes, further questions may prompt you to the answer of the question you were struggling with.

Tips regarding the questions

Multiple-choice questions
Below are tips on how to approach multiple-choice questions. The examples used are typical sample questions from Unit 016: Requirements of BS 7671.

Open Questions
These are questions where the answer completes the stem of the question to form a sentence. These are commonly used for open book exams relating to a particular quote or sentence within the book.

> Basic Protection can be provided by
>
> a. protective bonding conductors
> b. barriers and enclosures
> c. non-conducting locations
> d. class II equipment.

Closed Questions
These will have a stem that forms a clear question and an answer must be selected. These are the most commonly used questions overall.

> What is the maximum residual current rating of an RCD providing Additional Protection?
>
> a. 10 mA
> b. 30 mA
> c. 100 mA
> d. 300 mA

THE EXAM
TIPS REGARDING THE QUESTIONS

Negative questions
Although these types of questions are rare, they are the most common questions which are answered incorrectly by people who do not fully read the question. The question is asking you to find something that isn't correct using the word '**not**', which is usually in bold font.

> Which one of the following is **not** a method of Fault Protection?
>
> a. Electrical Separation
> b. Reinforced Insulation
> c. SELV and PELV
> d. Barriers/enclosures

Keywords
Look for keywords in a question, especially if it is an open book exam as the keywords should guide you to the correct location in the book. The keywords here are Basic Protection. Basic protection should guide you, in BS 7671, to Chapter 41; Protection against Electric Shock and in the contents page for Chapter 41, methods of Basic Protection are listed including Barriers and Enclosures. All the others are listed as Fault Protective Measures.

> Basic Protection can be provided by
>
> a. Protective bonding conductors
> b. Barriers and enclosures
> c. Non-conducting locations
> d. Class II equipment.

Notes

Notes

Written exams

Key terms

Questions will always begin with a key term. This is the indication of:

- how in depth your answer should be;
- how you present your answer;
- the level your answer needs to be at.
- Key terms often used are:
    - State – typically requires a one word answer or a short statement
    - List – usually more than three items needed (the number is often given)
    - Describe – often linked to a procedure such as, for an example, the isolation procedure
    - Explain – generally linked to a principle such as, for example, ohms law or electric shock
    - Calculate – requires maths
    - Determine – requires a mixture of calculations and references
    - Explain, with the aid of:
        - a diagram – requires a technical diagram, such as a circuit
        - a sketch – requires a drawing to show what something looks like

Keywords

Like multiple-choice, written exams will have key words that should guide you in the direction needed.

> State the three documents issued following an initial verification.

What are the keywords here?

three – three items needed

the – there are only three possible items. If there is no 'the', there could be a wider range of items.

documents – paper, books, journals, etc.

following – after, not before or during

initial verification – indicates a specific process

THE EXAM
PULL OUT THE INFORMATION FROM THE QUESTION OR SCENARIO

Pull out the information from the question or scenario

Some questions have large stems or scenarios giving key information that is often needed for questions that follow. It is always a good idea to read once, then read again highlighting or underlining the key information.

> A new electrical installation is to undergo an Initial Verification. The installation and supply form a 400 V three-phase TN-C-S system. At the origin of the installation, Z_s and PFC were measured and recorded as 0.3 ohms and 0.6 kA respectively. The temperature during the testing process is 20 °C. The final circuits are wired within PVC conduit.

> a) State the maximum test current for an RCD providing Additional Protection.
> b) List three instruments used for dead testing.
> c) State, for **each** answer given in b) above, **one** test where **each** instrument is used.

Some questions have a 'stem'. The stem contains the information relating to each question under it

Each question must relate to that scenario in the stem.

No stem then no relationship (unless stated)

Item a) has nothing to do with b) or c) but c) Links to b)

Note the key words **'each'** and **'one'** in c);

This means that one answer must be given to every answer you gave in b). Not one answer for all instruments given!

> An electrical installation within a large commercial building is to be inspected and tested on completion of the new works.
>
> State
>
> a) the tests required
> b) instruments needed for each test

Notes

Marks

The number of marks awarded for a question can be a good indicator to the depth of the answers you need to provide.

Presenting your answer

When presenting your answer, make sure it is neat and clear. Correct spelling, though desirable, isn't too essential providing it is easily understood and does not cause confusion.

Use bullet points if possible, especially if you are describing a procedure. Sometimes a diagram can speak a thousand words.

Always show calculations as marks are awarded to calculations and generally only one mark for the answer! You could get a calculation wrong but still get marks for the procedure.

Which of the following answers would you prefer to mark if you were the examiner?

Both of these answers to a question are the same and carry the same marks but if someone was marking Style A, they may miss a key point and therefore that candidate would lose marks.

Answer style A

I would get out of the van and open my test meter box. Selecting a low resistance ohmmeter I would fully isolate the circuit. I would then check my meter for function. I would then link line to cpc at the fuseboard and then disconnect all equipment, I would also bridge out all sensitive devices. I would then go to all the switches and lighting points on the circuit and test using my meter which does not need to be GS38 but I would null my leads before testing as it does make readings much more accurate. I would then take a reading at every switch and light then I would write down on the schedule of test results the biggest value I got using my low resistance ohmmeter.

Answer style B

- Fully isolate
- Use low resistance ohmmeter
- Null leads check meter
- Prepare circuit by removing or bridging sensitive devices
- Temporary link L-E at DB
- Test at all points
- Record highest as R1+R2
- Re-instate circuit

EXAM PRACTICE 1

Exam Practice 1

**Exam practice 1: Unit 101/001 understand health,
safety and environmental considerations**   26

**Exam practice 1 questions with answers: Unit 001
understand health, safety and environmental considerations**   32

**Exam practice 1 answer key: Unit 001 understand health,
safety and environmental considerations**   41

Exam practice 1: Unit 101/001 understand health, safety and environmental considerations

1. What equipment must an employer provide for employees who work on construction sites?

   - a. Company co-ordinated polo shirts.
   - b. Waterproof shoes or boots.
   - c. Goggles, gloves and hard hats.
   - d. Hand tools and tape measures.

2. Which of the following items require COSHH data/information to be available for users?

   - a. A painted set of stepladders.
   - b. PVC conduit adhesives.
   - c. Mobile platforms.
   - d. Pneumatic tools.

3. Which statutory regulation may have an impact on the time of day that electricians work on a large site within a residential area?

   - a. The Control of Noise at Work Regulations.
   - b. The Electricity at Work Regulations.
   - c. The Health and Safety at Work etc. Act.
   - d. The Management of Health and Safety at Work Regulations.

4. What materials or equipment do the WEEE Regulations apply to the disposal of?

   - a. Asbestos.
   - b. Chemicals.
   - c. Plastics.
   - d. Electronic.

EXAM PRACTICE 1

EXAM PRACTICE 1: UNIT 101/001 UNDERSTAND HEALTH, SAFETY AND ENVIRONMENTAL CONSIDERATIONS

5 What is the **most** appropriate initial action that should be taken when a work colleague falls into a deep trench and appears to be unconscious?

- ○ a. Call the emergency services.
- ○ b. Climb down and assess injuries.
- ○ c. Rope off the trench to secure it.
- ○ d. Shout loudly to try and revive them.

6 What is the first action that must be taken on discovering an unconscious person in contact with a live part?

- ○ a. Place them in the recovery position.
- ○ b. Check them for signs of burning.
- ○ c. Isolate the live part from the supply.
- ○ d. Insulate the person from the floor.

7 What immediate action must be taken if a water service pipe becomes damaged whilst digging a trench for a cable on a construction site?

- ○ a. Inform the site supervisor.
- ○ b. Repair the pipe with tape.
- ○ c. Place sand over the pipe.
- ○ d. Call the emergency services.

8 What is the **most** likely outcome if diesel fuel leaked into a surface water drain?

- ○ a. Pollution of local watercourses.
- ○ b. Explosion risk at nearby sewage works.
- ○ c. Contamination of local drinking water.
- ○ d. Fire risk in local toilets and sinks.

9 How should asbestos be removed from a building and disposed of?

- ○ a. By placing in an airtight bag in a skip.
- ○ b. By a specialist registered contractor.
- ○ c. By burying in a pit at least 3 m deep.
- ○ d. By burning it in an airtight container.

Notes

Notes

10 What is the **most** important reason for reporting a leak in an outdoor fuel storage tank?

- a. To minimise land contamination.
- b. To reduce the cost of lost fuel.
- c. To avoid bad odours from the fuel.
- d. To ensure a constant fuel supply.

11 When should risk assessments be reviewed?

- a. Every time the process changes.
- b. At the end of each working day.
- c. When an accident causes injury.
- d. On the first day of each month.

12 What should be the last form of risk reduction when a task creates large amounts of dust?

- a. Using water sprays.
- b. Using forced ventilation.
- c. Using dust extraction methods.
- d. Using personal protective equipment.

13 Which item of PPE is the **most** important when cutting a small hole into a plasterboard ceiling directly above?

- a. Hard hat.
- b. Hi-vis vest.
- c. Goggles.
- d. Boots.

14 When should a first aid box be present?

- a. When more than three people are working.
- b. They are not a mandatory requirement on sites.
- c. When more than twenty people are on site.
- d. They are mandatory at all places of work.

EXAM PRACTICE 1

EXAM PRACTICE 1: UNIT 101/001 UNDERSTAND HEALTH, SAFETY AND ENVIRONMENTAL CONSIDERATIONS

Notes

15 What is the risk of leaving an opened, unused sterile dressing in a first aid kit?

- a. The dressing will contaminate other sealed dressings.
- b. The dressing could cause infection if used on a wound.
- c. There is no risk if the dressing is rinsed before use.
- d. There is a risk of fire from fumes within the first aid box.

16 What is the correct ratio to give the correct gradient on a ladder against a wall?

- a. One metre up for every four metres out.
- b. Two metres up for every four metres out.
- c. Four metres up for every two metres out.
- d. Four metres up for every one metre out.

17 What is the **most** effective method of securing an isolated circuit on a busy construction site?

- a. Place a notice over the open circuit breaker.
- b. Remove the circuit breaker from the distribution board.
- c. Secure the circuit breaker using a lock with a unique key.
- d. Lock the circuit breaker using a lock with multiple keys.

18 What does the packaging warning sign in the image indicate?

- a. Explosive.
- b. Corrosive.
- c. Irritant.
- d. Oxidant.

Notes

19 What word is used to define something that has a potential to cause harm?

- a. Risk.
- b. Hazard.
- c. Accident.
- d. Injury.

20 Of the following, what is the **main** risk associated with trailing leads for power tools on construction sites?

- a. Shock due to damaged insulation.
- b. Fire due to overloading cables.
- c. Burns due to excessive voltages.
- d. Asphyxiation due to PVC fumes.

21 Which hazardous task could lead to a risk of burns?

- a. Connecting a cable to an electric heater.
- b. Measuring continuity of CPC at a light.
- c. Checking standby battery acid levels.
- d. Adjusting a room heating thermostat.

22 What action is the most appropriate method of risk reduction when electrical work needs to be carried out in an escape route in an office?

- a. Work fast to minimise blockage time.
- b. Carry out the work out of office hours.
- c. Decommission the fire alarm system.
- d. Barrier off the route at each end.

23 Which type of fire extinguisher is **most** suitable for electrical fires?

- a. CO_2.
- b. Water.
- c. Dry powder.
- d. Foam.

EXAM PRACTICE 1

EXAM PRACTICE 1: UNIT 101/001 UNDERSTAND HEALTH, SAFETY AND ENVIRONMENTAL CONSIDERATIONS

24 Which of the following could contain asbestos?

- a. PVC insulation.
- b. Foam pipe lagging.
- c. Felt roof sheeting.
- d. Artex ceilings.

25 What is the safest procedure if a cable is to be secured to a material where the presence of asbestos is suspected?

- a. Place PVC sheets over the material first.
- b. Keep fixing-hole diameters below 6 mm.
- c. Contact a specialist for advice and guidance.
- d. Dampen the material with water before contact.

Notes

Exam practice 1 questions with answers: Unit 001 understand health, safety and environmental considerations

1. What articles must an employer provide for employees who work on construction sites?

   - a) Company co-ordinated polo shirts.
   - b) Waterproof shoes or boots.
   - ● c) Goggles, gloves and hard hats.
   - d) Hand tools and tape measures.

   Correct answer c
   It is the duty of employers to provide the necessary PPE for their employees. Goggles, gloves and hard hats are the most commonly required items of PPE used on construction sites.

2. Which of the following items require COSHH data/information to be available for users?

   - a. A painted set of stepladders.
   - ● b. PVC conduit adhesives.
   - c. Mobile platforms.
   - d. Pneumatic tools.

   Correct answer b
   Control of Substances Hazardous to Health (COSHH). Regulations require users to be aware of the hazards when using chemical substances to avoid danger. The information should inform the user on correct storage, use and any actions should accidents, such as spillage, occur.

3. Which statutory regulation may have an impact on the time of day that electricians work on a large site within a residential area?

   - ● a. The Control of Noise at Work Regulations.
   - b. The Electricity at Work Regulations.
   - c. The Health and Safety at Work etc. Act.
   - d. The Management of Health and Safety at Work Regulations.

   Correct answer a
   Noise pollution from work activities may restrict the times that work is carried out on site, especially where they are in residential areas.

EXAM PRACTICE 1

EXAM PRACTICE 1 QUESTIONS WITH ANSWERS: UNIT 001 UNDERSTAND HEALTH, SAFETY AND ENVIRONMENTAL CONSIDERATIONS

Whilst the HSWA (option c) may indirectly affect the number of hours worked by operatives, it doesn't impact on the time of day activities take place.

4 What materials or equipment do the WEEE Regulations apply to the disposal of?

- a. Asbestos.
- b. Chemicals.
- c. Plastics.
- ● d. Electronic.

Correct answer d
WEEE Regulations is a commonly used abbreviation which stands for the Waste Electrical and Electronic Equipment Regulations which covers the disposal of electronic items of equipment.

5 What is the **most** appropriate initial action that should be taken when a work colleague falls into a deep trench and appears to be unconscious?

- ● a. Call the emergency services.
- b. Climb down and assess injuries.
- c. Rope off the trench to secure it.
- d. Shout loudly to try and revive them.

Correct answer a
In this situation you should never place yourself in danger. The emergency services are the best option to assess the risk of rescuing the injured person.

6 What is the first action that must be taken on discovering an unconscious person in contact with a live part?

- a. Place them in the recovery position.
- b. Check them for signs of burning.
- ● c. Isolate the live part from the supply.
- d. Insulate the person from the floor.

Correct answer c
Before a person is treated, in this situation, the situation needs to be made safe before anybody touches the person and receives an electric shock themselves. Isolating the live part from the supply will then enable treatment to be carried out and remove the immediate danger to the patient.

Notes

7. What immediate action must be taken if a water service pipe becomes damaged whilst digging a trench for a cable on a construction site?

   - ◉ a. Inform the site supervisor.
   - ○ b. Repair the pipe with tape.
   - ○ c. Place sand over the pipe.
   - ○ d. Call the emergency services.

   Correct answer a
   As the incident is beyond the responsibility of the person installing the cable, the site supervisor must be informed immediately so they can take appropriate action. Attempts to repair the pipe should never be undertaken as contamination of the water supply, amongst other reasons, may be a risk.

8. What is the **most** likely outcome if diesel fuel leaked into a surface water drain?

   - ◉ a. Pollution of local watercourses.
   - ○ b. Explosion risk at nearby sewage works.
   - ○ c. Contamination of local drinking water.
   - ○ d. Fire risk in local toilets and sinks.

   Correct answer a
   Surface water drains are generally linked to local watercourses such as streams and rivers, not sewage systems and certainly not drinking water. Spillage of chemicals or fuels into surface drains will lead to pollution of the local rivers and streams.

9. How should asbestos be removed from a building and disposed of?

   - ○ a. By placing in an airtight bag in a skip.
   - ◉ b. By a specialist registered contractor.
   - ○ c. By burying in a pit at least 3 m deep.
   - ○ d. By burning it in an airtight container.

   Correct answer b
   Asbestos should only be handled and removed for disposal by registered specialist contractors. Disturbing asbestos can be extremely hazardous to health if the correct procedures are not followed.

EXAM PRACTICE 1

EXAM PRACTICE 1 QUESTIONS WITH ANSWERS: UNIT 001 UNDERSTAND HEALTH, SAFETY AND ENVIRONMENTAL CONSIDERATIONS

Notes

10 What is the **most** important reason for reporting a leak in an outdoor fuel storage tank?

- ◉ a. To minimise land contamination.
- ○ b. To reduce the cost of lost fuel.
- ○ c. To avoid bad odours from the fuel.
- ○ d. To ensure a constant fuel supply.

Correct answer a

Although leaking fuel may be costly and produce bad smells, the impact of land contamination from the fuel will have the most severe consequences to the environment. In addition, if unreported, agencies or authorities could prosecute, leading to fines that are greater than the cost of lost fuel.

11 When should risk assessments be reviewed?

- ◉ a. Every time the process changes.
- ○ b. At the end of each working day.
- ○ c. When an accident causes injury.
- ○ d. On the first day of each month.

Correct answer a

Many organisations use generic risk assessments for tasks or processes. Each time the process changes, the risk changes and must therefore be reassessed.

12 What should be the last form of risk reduction when a task creates large amounts of dust?

- ○ a. Using water sprays.
- ○ b. Using forced ventilation.
- ○ c. Using dust extraction methods.
- ◉ d. Using personal protective equipment.

Correct answer d

PPE should always be the last form of risk reduction.

Notes

13 Which item of PPE is the **most** important when cutting a small hole into a plasterboard ceiling directly above?

- a. Hard hat.
- b. Hi-vis vest.
- **c. Goggles.** ●
- d. Boots.

Correct answer c
The greatest risk to a person cutting into plasterboard is dust in the eyes. As the ceiling is directly above, it is unlikely that falling objects are a likely risk or movement of traffic. Although all the items listed are standard PPE, it is the goggles that are most important for this task.

14 When should a first aid box be present?

- a. When more than three people are working.
- b. They are not a mandatory requirement on sites.
- c. When more than twenty people are on site.
- **d. They are mandatory at all places of work.** ●

Correct answer d
Basic first aid facilities must be available at all times at places of work regardless of the number of people working.

15 What is the risk of leaving an opened, unused sterile dressing in a first aid kit?

- a. The dressing will contaminate other sealed dressings.
- **b. The dressing could cause infection if used on a wound.** ●
- c. There is no risk if the dressing is rinsed before use.
- d. There is a risk of fire from fumes within the first aid box.

Correct answer b
Packaging containing any sterile equipment must always be disposed of if opened as contamination could cause serious infection if used to treat a person.

EXAM PRACTICE 1

EXAM PRACTICE 1 QUESTIONS WITH ANSWERS: UNIT 001 UNDERSTAND HEALTH, SAFETY AND ENVIRONMENTAL CONSIDERATIONS

Notes

16 What is the correct ratio to give the correct gradient on a ladder against a wall?

- a. One metre up for every four metres out.
- b. Two metres up for every four metres out.
- c. Four metres up for every two metres out.
- ● d. Four metres up for every one metre out.

Correct answer d
The correct ratio when erecting a ladder is, for every four metres up, the ladder must be one metre out from the wall to give the correct safe and stable gradient.

17 What is the **most** effective method of securing an isolated circuit on a busy construction site?

- a. Place a notice over the open circuit breaker.
- b. Remove the circuit breaker from the distribution board.
- ● c. Secure the circuit breaker using a lock with a unique key.
- d. Lock the circuit breaker using a lock with multiple keys.

Correct answer c
Following isolation, the circuit breaker controlling the circuit should be secured with a lock having one unique key which the electrician should hold at all times. This ensures that the circuit cannot be energised without the knowledge of the person working on it.

Notes

18 What does the packaging warning sign in the image indicate?

- ● a. Explosive.
- ○ b. Corrosive.
- ○ c. Irritant.
- ○ d. Oxidant.

Correct answer a
The Chemical (Hazard Information and Packaging for Supply) Regulations (CHIP) uses the image shown to indicate that a substance is explosive.

19 What word is used to define something that has a potential to cause harm?

- ○ a. Risk.
- ● b. Hazard.
- ○ c. Accident.
- ○ d. Injury.

Correct answer b
Something that has a potential to cause harm is a hazard. The likelihood that harm will happen is the level of risk from the hazard.

20 Of the following, what is the **main** risk associated with trailing leads for power tools on construction sites?

- ● a. Shock due to damaged insulation.
- ○ b. Fire due to overloading cables.
- ○ c. Burns due to excessive voltages.
- ○ d. Asphyxiation due to PVC fumes.

Correct answer a
One of the greatest risks from the use of trailing extension leads on construction sites is electric shock due to damage of the cable insulation due to site conditions.

EXAM PRACTICE 1

EXAM PRACTICE 1 QUESTIONS WITH ANSWERS: UNIT 001 UNDERSTAND HEALTH, SAFETY AND ENVIRONMENTAL CONSIDERATIONS

Notes

21 Which hazardous task could lead to a risk of burns?

- a. Connecting a cable to an electric heater.
- b. Measuring continuity of CPC at a light.
- ● c. Checking standby battery acid levels.
- d. Adjusting a room heating thermostat.

Correct answer c
Battery acid is hazardous and if handled incorrectly, could burn the skin if a small amount is spilt. Whilst heaters and lights are hazards, the situations described in the question would mean the equipment is not posing a risk at the time.

22 What action is the most appropriate method of risk reduction when electrical work needs to be carried out in an escape route in an office?

- a. Work fast to minimise blockage time.
- ● b. Carry out the work out of office hours.
- c. Decommission the fire alarm system.
- d. Barrier off the route at each end.

Correct answer b
It is critical for the safety of people working in the office that escape routes are clear of any obstruction whilst the building is occupied. The most effective way to ensure this is to carry out work whilst the building is empty such as over evenings or weekends.

23 Which type of fire extinguisher is **most** suitable for electrical fires?

- ● a. CO_2.
- b. Water.
- c. Dry powder.
- d. Foam.

Correct answer a
CO_2 is a non-conductive gas and is the most suitable for tackling fires to electrical equipment which could be live.

Notes

24 Which of the following could contain asbestos?

- a. PVC insulation.
- b. Foam pipe lagging.
- c. Felt roof sheeting.
- ● d. Artex ceilings.

Correct answer d
Many older properties contain asbestos in the ceiling materials and should never be disturbed as the dust particles are very hazardous.

25 What is the safest procedure if a cable is to be secured to a material where the presence of asbestos is suspected?

- a. Place PVC sheets over the material first.
- b. Keep fixing-hole diameters below 6 mm.
- ● c. Contact a specialist for advice and guidance.
- d. Dampen the material with water before contact.

Correct answer c
When dealing with any materials where asbestos is suspected, a specialist contractor must be consulted before any work is undertaken to assess the best methods for dealing with the material.

Exam practice 1 answer key: Unit 001 understand health, safety and environmental considerations

1 c	6 c	11 a	16 d	21 c
2 b	7 a	12 d	17 c	22 b
3 a	8 a	13 c	18 a	23 a
4 d	9 b	14 d	19 b	24 d
5 a	10 a	15 b	20 a	25 c

Notes

EXAM PRACTICE 2

Exam Practice 2

Exam practice 2: Unit 003 electrical scientific principles and technologies — 44

Exam practice 2 questions with answers: Unit 003 electrical scientific principles and technology — 56

Exam practice 2 answer key: Unit 003 electrical scientific principles and technologies — 76

Notes

Exam practice 2: Unit 003 electrical scientific principles and technologies

1 What is 75% of 40?

  ○ a. 18
  ○ b. 24
  ○ c. 30
  ○ d. 53

2 What is an alternative way to write the value of 18 million?

  ○ a. 18×10^{-6}
  ○ b. 18×10^{-3}
  ○ c. 18×10^{3}
  ○ d. 18×10^{6}

3 What would be measured using the SI unit m^3?

  ○ a. Length.
  ○ b. Volume.
  ○ c. Area.
  ○ d. Mass.

4 What is the SI unit used to measure temperature?

  ○ a. Kelvin.
  ○ b. Celsius.
  ○ c. Joule.
  ○ d. Fahrenheit.

5 What electrical unit is used to measure resistance?

  ○ a. Watt.
  ○ b. Volt.
  ○ c. Ampere.
  ○ d. Ohm.

EXAM PRACTICE 2

EXAM PRACTICE 2: UNIT 003 ELECTRICAL SCIENTIFIC PRINCIPLES AND TECHNOLOGIES

Notes

6 What unit is used to measure capacitance?

- a. Henry.
- b. Joule.
- c. Farad.
- d. Watt.

7 What instrument, used by electricity supply companies, measures the electrical energy used in a house?

- a. Wattmeter.
- b. kWh meter.
- c. Voltmeter.
- d. Ammeter.

8 Which of the following, when applied to the mass of an object, gives the object weight?

- a. Gravity.
- b. Density.
- c. Resistivity.
- d. Energy.

9

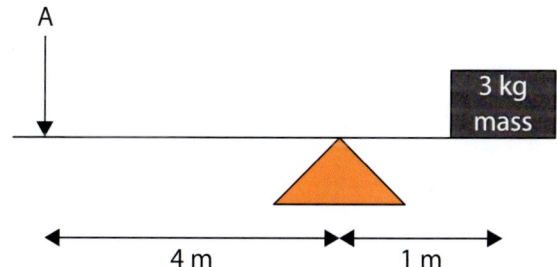

How much mass, when placed at point A in the image, would keep the lever perfectly balanced?

- a. 0.75 kg.
- b. 1.00 kg.
- c. 2.75 kg.
- d. 4.00 kg.

10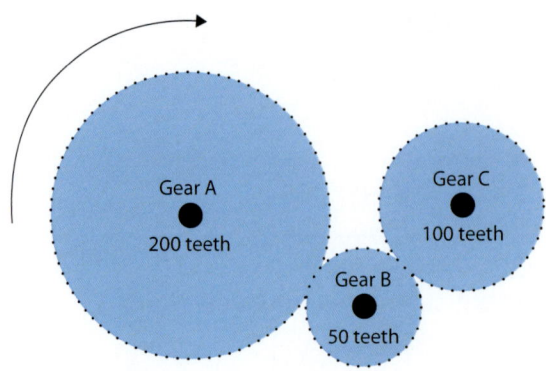

Gear A rotates at a rate of 1 revolution per minute in the direction shown in the above image. How many times per minute would gear B rotate?

- a. 1
- b. 2
- c. 3
- d. 4

11 What is calculated using the following formula?

$$\frac{force \times distance}{time}$$

- a. Mechanical power in Watts.
- b. Energy in Joules.
- c. The mass of an object in kg.
- d. The efficiency of a machine.

12 What is the output power of a machine that is 80% efficient and has an input power of 120 W?

- a. 40 W
- b. 78 W
- c. 96 W
- d. 150 W

EXAM PRACTICE 2

EXAM PRACTICE 2: UNIT 003 ELECTRICAL SCIENTIFIC PRINCIPLES AND TECHNOLOGIES

13 What is the amount of work required, in joules, to raise a force of 20 N a distance of 30 m?

- a. 1.5 J.
- b. 10 J.
- c. 60 J.
- d. 600 J.

14 How much power is required to raise a force of 10 N a distance of 2 m in 15 seconds?

- a. 1.3 W.
- b. 75 W.
- c. 300 W.
- d. 600 W.

15

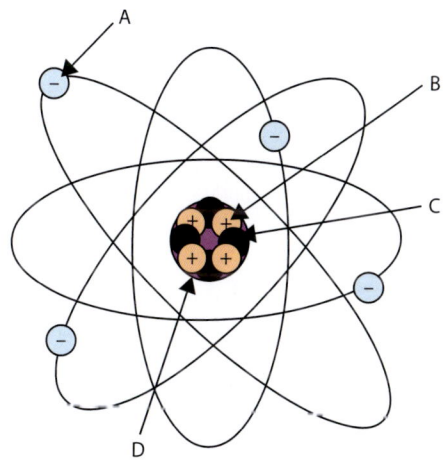

What is the atomic part labelled A in the image above?

- a. Proton.
- b. Electron.
- c. Neutron.
- d. Nucleus.

16 What would be a good insulator at low voltages?

- a. Brass.
- b. Glass.
- c. Gold.
- d. Copper.

Notes

Notes

17 What is the resistance of a 20 m long conductor, at 20 °C, having a cross-sectional area of 2.5 mm² and a resistivity of 0.0172 µΩ/m³?

- a. 0.002 Ω
- b. 0.137 Ω
- c. 0.860 Ω
- d. 1.003 Ω

18 What two factors affect the resistance of a material?

- a. Length and weight.
- b. Cross-sectional area and weight.
- c. Weight and gravity.
- d. Cross-sectional area and length.

19

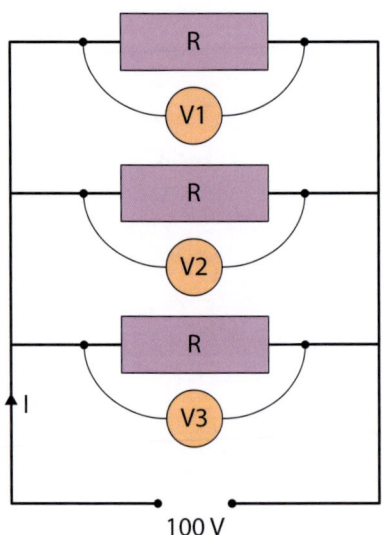

What is the relationship between voltage and current at the resistors for the circuit shown above?

- a. Different voltage, same current at each resistor.
- b. Different voltage, different current at each resistor.
- c. Same voltage, same current at each resistor.
- d. Same voltage, different current at each resistor.

EXAM PRACTICE 2

EXAM PRACTICE 2: UNIT 003 ELECTRICAL SCIENTIFIC PRINCIPLES AND TECHNOLOGIES

20 What type of circuit would cause the voltage to vary over **each** resistor in the circuit but the current would remain constant?

- a. Parallel.
- b. Series.
- c. Series/parallel.
- d. Parallel/series.

21

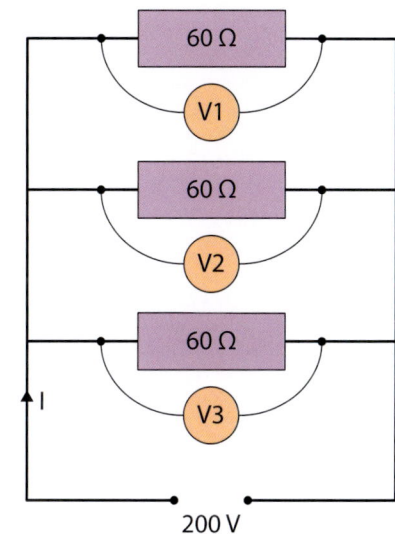

What is the value of voltage measured by V1 in the diagram above?

- a. 20 V
- b. 30 V
- c. 200 V
- d. 600 V

22

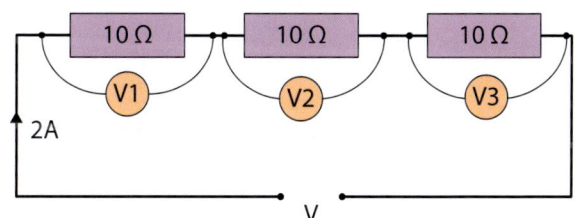

What is the value of voltage measured by V3 in the diagram above?

- a. 60 V
- b. 20 V
- c. 5 V
- d. 2 V

23

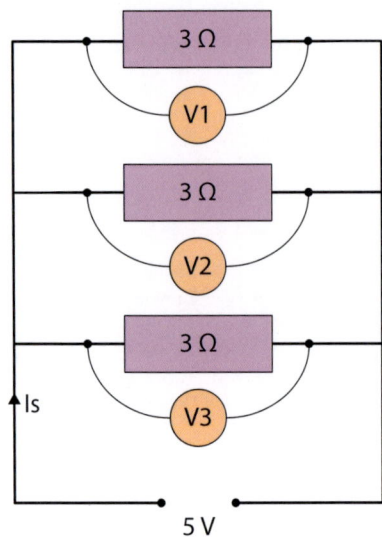

What would be the supply current(s) for the circuit shown in the diagram above ?

- a. 0.55 A
- b. 1.00 A
- c. 5.00 A
- d. 9.00 A

24

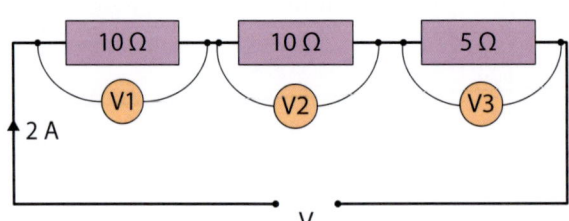

What is the value of current flowing through the 5 Ω resistor in the diagram shown above?

- a. 2 A
- b. 3 A
- c. 5 A
- d. 25 A

EXAM PRACTICE 2

EXAM PRACTICE 2: UNIT 003 ELECTRICAL SCIENTIFIC PRINCIPLES AND TECHNOLOGIES

25 What is the total resistance of a circuit which has four resistors connected in series, where each resistor has a value of 10 Ω?

- a. 2.5 Ω
- b. 4.0 Ω
- c. 10 Ω
- d. 40 Ω

26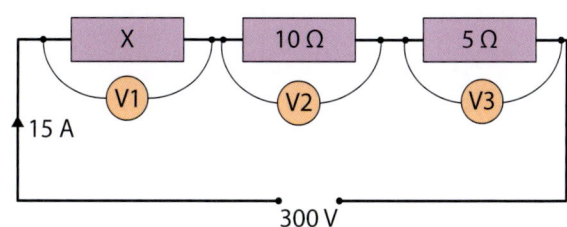

What is the value of the resistor marked X in the diagram shown above?

- a. 15 Ω
- b. 10 Ω
- c. 5 Ω
- d. 2 Ω

27 What is the total power in a DC circuit having a total resistance of 30 Ω and a supply current of 4 A?

- a. 7.5 W
- b. 120 W
- c. 480 W
- d. 3600 W

28 What term describes the loss of voltage, due to circuit resistance, between the supply and load in a circuit?

- a. Voltage drop.
- b. Current gain.
- c. Power slip.
- d. Resistance dip.

Notes

Notes

29 What will happen to a conductor if the circuit current is significantly increased?

- a. The conductor's temperature will reduce.
- b. The conductor's temperature will increase.
- c. The conductor's cross-sectional area will decrease.
- d. The conductor's length will decrease.

30

Magnet A   Magnet B

What will happen when the two bar magnets are placed together as shown in the image above?

- a. Magnets A and B will rotate on their axis like a motor.
- b. Magnet A will move upwards and magnet B downwards.
- c. Magnets A and B will attract and form one larger magnet.
- d. Magnet A will be pushed to the left and magnet B to the right.

31 What is magnetic flux density?

- a. The amount of flux emitted by a magnet measured in Webers.
- b. The amount of flux emitted by a magnet measured in Teslas.
- c. The amount of flux in a given area measured in Webers.
- d. The amount of flux in a given area measured in Teslas.

32 Which image shows the correct direction of magnetic flux flow for the current carrying conductor?

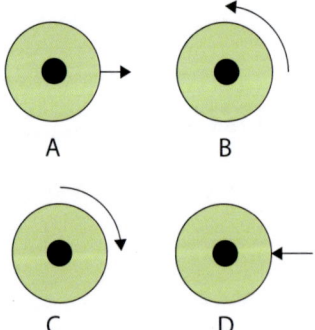

- a. A
- b. B
- c. C
- d. D

33 Which point of rotation would the maximum emf be induced by the simple alternator shown in the image below?

- a. A
- b. B
- c. C
- d. D

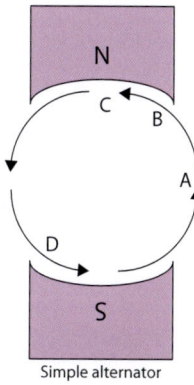
Simple alternator

34 What is the periodic time for a sinewave indicating a frequency of 50 Hz?

- a. 5 ms
- b. 20 ms
- c. 30 ms
- d. 50 ms

35 What would be the emf induced from a simple alternator, having a 0.1 m conductor, rotating at a velocity of 40 metres per second within a magnetic field having a flux density of 4 Tesla?

- a. 16 V
- b. 50 V
- c. 100 V
- d. 230 V

Notes

36 Which quantity, indicated on the sinewave in the image below, would be multiplied by 0.707 to obtain the r.m.s. voltage?

○ a. A
○ b. B
○ c. C
○ d. D

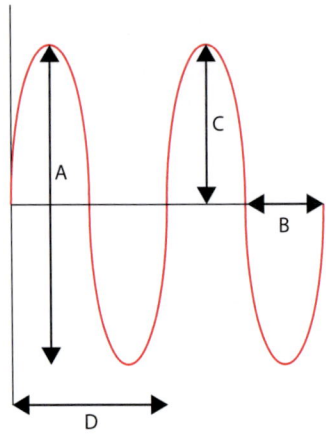

37 What single electronic component provides half wave rectification?

○ a. Capacitor.
○ b. Inductor.
○ c. Diode.
○ d. Transistor.

38 What electronic component would be used to measure the water temperature within a central heating boiler?

○ a. Thermistor.
○ b. Invertor.
○ c. Resistor.
○ d. Transistor.

39 What electronic component stores a charge?

○ a. Inductor.
○ b. Transistor.
○ c. Capacitor.
○ d. Resistor.

EXAM PRACTICE 2
EXAM PRACTICE 2: UNIT 003 ELECTRICAL SCIENTIFIC PRINCIPLES AND TECHNOLOGIES

40 What electronic component is commonly used as a light source or indicator?

- a. SCR
- a. PCP
- b. LED
- c. LDD

Notes

Notes

Exam practice 2 questions with answers: Unit 003 electrical scientific principles and technology

1 What is 75% of 40?

- a. 18
- b. 24
- ● c. 30
- d. 53

Correct answer c
The correct formula is:

$$\frac{total \times \%}{100} = value$$

so

$$\frac{40 \times 75}{100} = 30$$

2 What is an alternative way to write the value of 18 million?

- a. 18×10^{-6}
- b. 18×10^{-3}
- c. 18×10^{3}
- ● d. 18×10^{6}

Correct answer d
Where indices are negative, or to the power of minus, they are decimal values so these ones are not correct. A good method to remember the indices is to think of how many zeros are in the value. A million has six zeros, so 18 million would be 18×10^6. The other values are:

$18 \times 10^{-6} = 0.000018$

$18 \times 10^{-3} = 0.018$

$18 \times 10^{3} = 18000$

Tip - be familiar with your `ENG` **and** `x10^x` **functions on your calculator**

EXAM PRACTICE 2

EXAM PRACTICE 2 QUESTIONS WITH ANSWERS: UNIT 003 ELECTRICAL SCIENTIFIC PRINCIPLES AND TECHNOLOGY

Notes

3 What would be measured using the SI unit m^3?

- a. Length.
- ● b. Volume.
- c. Area.
- d. Mass.

Correct answer b

Volume is the measurement of a three-dimensional space, so would be measured in m^3.

Length is the distance from one point to another so would be measured in metres.

Area is the measurement of a flat space using two dimensions so would be measured in m^2.

Mass is measured in kg not metres.

4 What is the SI unit used to measure temperature?

- ● a. Kelvin.
- b. Celsius.
- c. Joule.
- d. Fahrenheit.

Correct answer a

Although a commonly used unit is Celsius, it isn't the official SI unit. Fahrenheit is a historical alternative to Celsius and still used in weather forecasts. Kelvin is the official SI unit because 'absolute zero' or zero Kelvin is the temperature where all matter freezes, not just water.

5 What electrical unit is used to measure resistance?

- a. Watts.
- b. Volts.
- c. Amperes.
- ● d. Ohms.

Correct answer d

The unit used to measure the resistance of a conductor, or indeed an insulator is the ohm (Ω).

Notes

6 What unit is used to measure capacitance?

- ○ a. Henry.
- ○ b. Joule.
- ● c. Farad.
- ○ d. Watt.

Correct answer c
The **Farad** is the base unit to measure capacitance. The **Henry** is used for inductance, **Joule** is energy and the **Watt** is power.

7 What instrument, used by electricity supply companies, measures the electrical energy used in a house?

- ○ a. Wattmeter.
- ● b. kWh meter.
- ○ c. Voltmeter.
- ○ d. Ammeter.

Correct answer b
Household electricity is metered on how much power is used over a period of time, which is energy, so the meter measures kilo watts per hour (kWh).

8 Which of the following, when applied to the mass of an object, gives the object weight?

- ● a. Gravity.
- ○ b. Density.
- ○ c. Resistivity.
- ○ d. Energy.

Correct answer a
An object having a mass would be weightless in space where there is zero gravity so the more weight increases the more gravity has an effect.

9

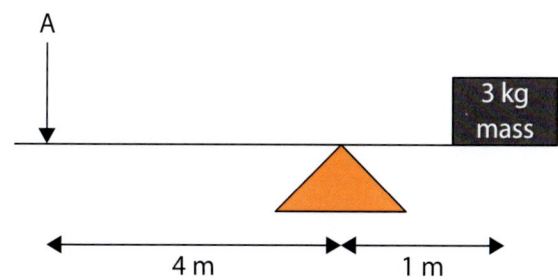

How much mass, when placed at point A in the image, would keep the lever perfectly balanced?

- ⦿ a. 0.75 kg.
- ○ b. 1.00 kg.
- ○ c. 2.75 kg.
- ○ d. 4.00 kg.

Correct answer a

Using the formula

$$force \times distance = force \times distance$$

and inserting known values

$$A \times 4\ m = 1\ m \times 3\ kg$$

then transposing to find the value of A

$$A = \frac{1\ m \times 3\ kg}{4\ m}$$

10

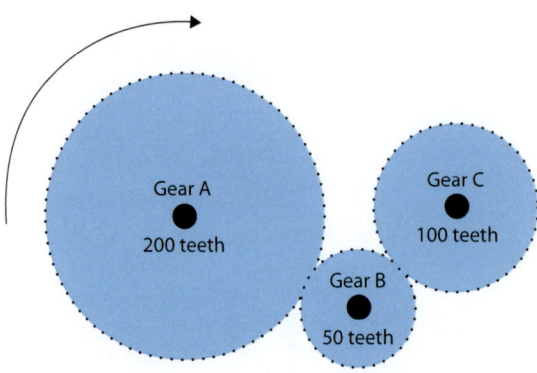

Gear A rotates at a rate of 1 revolution per minute in the direction shown in the image. How many times per minute would gear B rotate?

- a. 1
- b. 2
- c. 3
- ● d. 4

Correct answer d
As gear A has four times more teeth than gear B, then gear B would rotate four times more than gear A which is four times per minute.

Make sure you read the question fully as you may be tempted to work out the rotation of Gear C as it is also in the image but the question does not ask for this!

11 What is calculated using the following formula?

$$\frac{force \times distance}{time}$$

- ● a. Mechanical power in Watts.
- b. Energy in Joules.
- c. The mass of an object in kg.
- d. The efficiency of a machine.

Correct answer a
The formula shown is used to calculate the power needed to move an object against a force (of mass x gravity) for a particular distance over a period of time.

EXAM PRACTICE 2

EXAM PRACTICE 2 QUESTIONS WITH ANSWERS: UNIT 003 ELECTRICAL SCIENTIFIC PRINCIPLES AND TECHNOLOGY

12 What is the output power of a machine that is 80% efficient and has an input power of 120 W?

- a. 40 W
- b. 78 W
- ● c. 96 W
- d. 150 W

Correct answer c

As

$$\text{efficiency (\%)} = \frac{\text{power out} \times 100}{\text{power in}}$$

Then

$$\text{power out} = \frac{\% \times \text{power in}}{100}$$

So

$$\text{power out} = \frac{80 \times 120}{100} = 96\ W$$

13 What is the amount of work required, in joules, to raise a force of 20 N a distance of 30 m?

- a. 1.5 J
- b. 10 J
- c. 60 J
- ● d. 600 J

Correct answer d

As

$$\text{work} = \text{force} \times \text{distance}$$

Then

$$\text{work} = 20\ N \times 30\ m = 600\ J$$

Notes

Notes

14 How much power is required to raise a force of 10 N a distance of 2 m in 15 seconds?

- ◉ a. 1.3 W
- ○ b. 75 W
- ○ c. 300 W
- ○ d. 600 W

Correct answer a

As

$$power = \frac{force \times distance}{time}$$

Then

$$\frac{10 \times 2}{15} = 1.3\,W$$

Tip: As all of the values given are in the base SI units, no conversions are needed. Sometimes the values given may be in derived units. For example, the time may be given in minutes. In that situation, you must divide the time by 60 to convert into seconds.

15

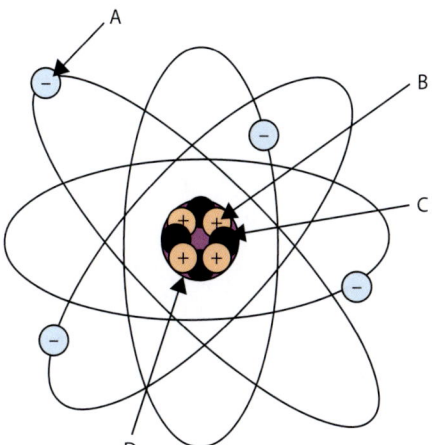

What is the atomic part labelled A in the image?

- a. Proton.
- **b. Electron.** ●
- c. Neutron.
- d. Nucleus.

Correct answer b
Within an atom, the orbiting part is called the electron. Electricity is the flow of electrons from one atom to another in a uniform direction. When a particular amount of electrons flow in a given direction, it is known as charge. The amount of charge over a given time is the amount of amperes.

16 What would be a good insulator at low voltages?

- a. Brass.
- **b. Glass.** ●
- c. Gold.
- d. Copper.

Correct answer b
Glass does not allow the flow of current at low voltages so is classed as an insulator.

All the other materials listed in the question are all good conductors and are commonly used to conduct current in electrical circuits and equipment.

The reason the voltage is mentioned in the question is because many materials which may be classed as insulators will break down if the voltage level is increased.

Notes

17 What is the resistance of a 20 m long conductor, at 20 °C, having a cross-sectional area of 2.5 mm² and a resistivity of 0.0172 µΩ/m³?

- a. 0.002 Ω
- **b. 0.137 Ω** ●
- c. 0.860 Ω
- d. 1.003 Ω

Correct answer b
As the resistance of a conductor is calculated, using resistivity, by

$$R = \frac{\rho L}{A} \text{ in ohms}$$

Then

$$R = \frac{0.0172 \times 10^{-6} \times 20}{2.5 \times 10^{-6}}$$

(Note: as the cross-sectional area is mm² not just mm, it is at x10^{-6} instead of x10^{-3}. Resistivity is x10^{-6} as it is given in micro ohms)

As the two sets of x10^{-6} cancel each other out, then

$$\frac{0.0172 \times 20}{2.5} = 0.137 \text{ Ω}$$

Tip: Be very careful as the incorrect answers, known as distractors, could very well be values where the formula is in the incorrect order.

18 What two factors affect the resistance of a material?

- a. Length and weight.
- b. Cross-sectional area and weight.
- c. Weight and gravity.
- **d. Cross-sectional area and length.** ●

Correct answer d
Every material which conducts electricity has a resistance. The resistance will increase with an increase in length but decrease with an increase in cross-sectional area.

EXAM PRACTICE 2

EXAM PRACTICE 2 QUESTIONS WITH ANSWERS: UNIT 003 ELECTRICAL SCIENTIFIC PRINCIPLES AND TECHNOLOGY

19

[Circuit diagram showing three resistors R in parallel, each with voltmeters V1, V2, V3 across them, connected to a 100 V supply with current I.]

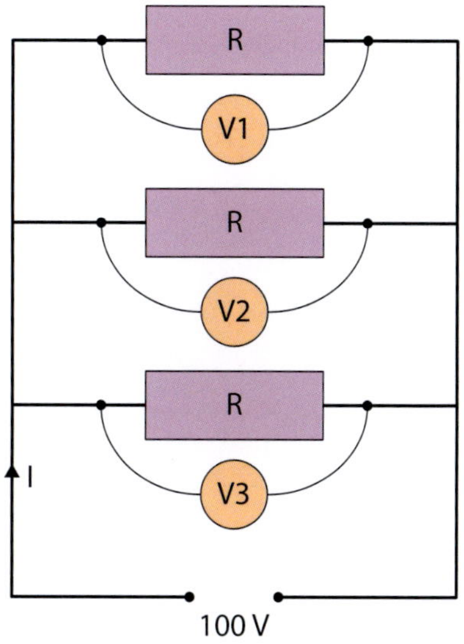

What is the relationship between voltage and current at the resistors for the circuit shown above?

- ○ a. Different voltage, same current at each resistor.
- ○ b. Different voltage, different current at each resistor.
- ○ c. Same voltage, same current at each resistor.
- ● d. Same voltage, different current at each resistor.

Correct answer d
The circuit shown is a parallel circuit where the voltage is equal across each resistor but the total circuit current is split over the resistors depending on their value.

20 **What type of circuit would cause the voltage to vary over each resistor in the circuit but the current would remain constant?**

- ○ a. Parallel.
- ● b. Series.
- ○ c. Series/parallel.
- ○ d. Parallel/series.

Correct answer b
In a series circuit, the circuit current would be the same through each resistor but the voltage across each resistor would be different depending on the resistor value.

Notes

21

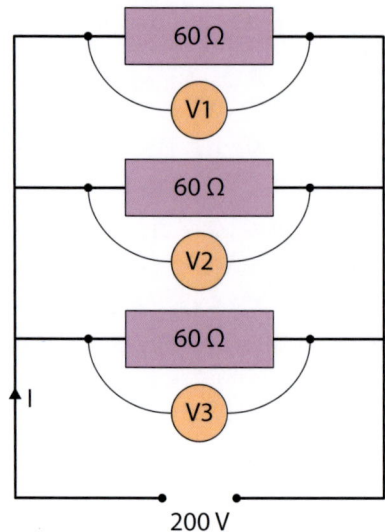

What is the value of voltage measured by V1 in the image?

- a. 20 V
- b. 30 V
- c. 200 V
- d. 600 V

Correct answer c

In a parallel circuit, such as the one shown, the voltage across each resistance is equal to the supply voltage (assuming no losses in the cables supplying the resistors).

22

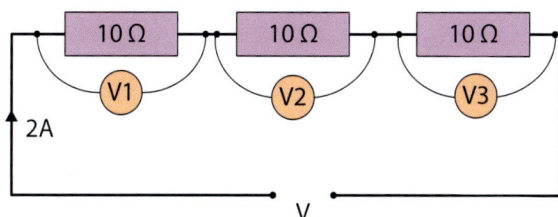

What is the value of voltage measured by V3 in the image?

- a. 60 V
- **b. 20 V**
- c. 5 V
- d. 2 V

Correct answer b

In the series circuit shown, the current is constant through each resistor so the current flowing through the 10 Ω resistor measured by V3 is 2 A. Applying Ohm's law to determine the voltage across that resistor

$$V = I \times R$$

So

$$V = 2 \times 10 = 20 \ V$$

Notes

23

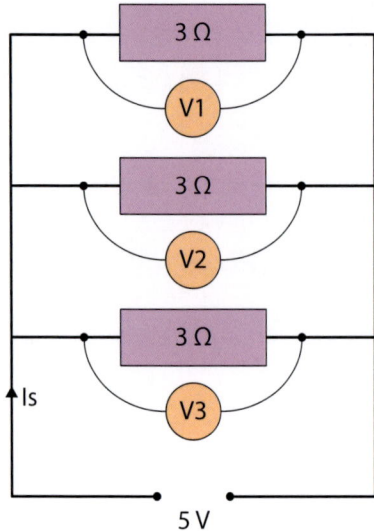

What would be the supply current(s) for the circuit shown in the image?

- a. 0.55 A
- b. 1.00 A
- ● c. 5.00 A
- d. 9.00 A

Correct answer c

The supply current (Is) is calculated using Ohm's law based on the supply voltage and the total circuit resistance.

In order to determine the supply current, the total circuit resistance must first be calculated.

As the image shows a parallel circuit, then the total resistance is found by

$$\frac{1}{R_t} = \frac{1}{R_1} + \frac{1}{R_2} + \frac{1}{R_3}$$

So

$$\frac{1}{R_t} = \frac{1}{3} + \frac{1}{3} + \frac{1}{3} = 1\,\Omega$$

Then applying Ohm's law

$$I = \frac{V}{R} \text{ so } \frac{5}{1} = 5\,A$$

EXAM PRACTICE 2

EXAM PRACTICE 2 QUESTIONS WITH ANSWERS: UNIT 003 ELECTRICAL SCIENTIFIC PRINCIPLES AND TECHNOLOGY

24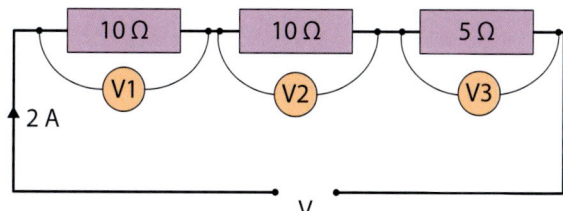

What is the value of current flowing through the 5 Ω resistor?

- ◉ a. 2 A
- ○ b. 3 A
- ○ c. 5 A
- ○ d. 25 A

Correct answer a

In a series circuit, the current is constant so the supply current of 2 A cannot change as it flows through the circuit.

25 What is the total resistance of a circuit which has four resistors connected in series, where each resistor has a value of 10 Ω?

- ○ a. 2.5 Ω
- ○ b. 4.0 Ω
- ○ c. 10 Ω
- ◉ d. 40 Ω

Correct answer d

In a series circuit, the total resistance is calculated by adding all the resistances together so

$$R_t = R_1 + R_2 + R_3 + R_4$$

So

$$R_t = 10 + 10 + 10 + 10 = 40 \, \Omega$$

26

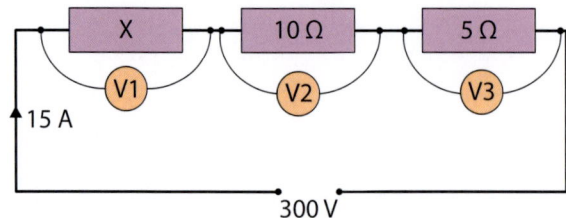

What is the value of the resistor marked X in the image?

- a. 15 Ω
- b. 10 Ω
- **c. 5 Ω** ✓
- d. 2 Ω

Correct answer c
To find the value of the resistor marked X in the image, the total resistance first needs to be calculated using the supply voltage and current using Ohm's law so

$$R_t = \frac{V}{I} \text{ so } \frac{300}{15} = 20\,\Omega$$

If the total resistance is 20 Ω and the known resistors total 15 Ω then, being a series circuit, resistor X is

$$20\,\Omega - 15\,\Omega = 5\,\Omega$$

27 What is the total power dissipated in a DC circuit having a total resistance of 30 Ω and a supply current of 4 A?

- a. 7.5 W
- b. 120 W
- **c. 480 W** ✓
- d. 3600 W

Correct answer c
In order to calculate the total power, given the information contained in the question, the formula to use is

$$P = I^2 R$$

So

$$P = 4^2 \times 30 = 480\ W$$

EXAM PRACTICE 2

EXAM PRACTICE 2 QUESTIONS WITH ANSWERS: UNIT 003 ELECTRICAL SCIENTIFIC PRINCIPLES AND TECHNOLOGY

Notes

28 What term describes the loss of voltage, due to circuit resistance, between the supply and load in a circuit?

- ◉ a. Voltage drop.
- ○ b. Current gain.
- ○ c. Power slip.
- ○ d. Resistance dip.

Correct answer a
Where the voltage measured at the terminals of a load is less than the supply voltage, voltage drop has occurred. The amount of voltage dropped or lost is affected by the resistance of the circuit conductor and the load current flowing in the circuit.

29 What will happen to a conductor if the circuit current is significantly increased?

- ○ a. The conductor's temperature will reduce.
- ◉ b. The conductor's temperature will increase.
- ○ c. The conductor's cross-sectional area will decrease.
- ○ d. The conductor's length will decrease.

Correct answer b
Increasing the current flowing in a conductor will increase the conductor's temperature.

30

Magnet A   Magnet B

What will happen when the two bar magnets are placed together as shown in the image?

- ○ a. Magnets A and B will rotate on their axis like a motor.
- ○ b. Magnet A will move upwards and magnet B downwards.
- ○ c. Magnets A and B will attract and form one larger magnet.
- ◉ d. Magnet A will be pushed to the left and magnet B to the right.

Correct answer d
As opposites attract, likes repel. As the two North poles are placed together, the two magnets will repel pushing away from each other.

Notes

31 What is magnetic flux density?

- ○ a. The amount of flux emitted by a magnet measured in Webers.
- ○ b. The amount of flux emitted by a magnet measured in Teslas.
- ○ c. The amount of flux in a given area measured in Webers.
- ● d. The amount of flux in a given area measured in Teslas.

Correct answer d
The amount of flux emitted by a magnet is simply the measure of flux and this is measured in Webers.

However, the amount of flux within a given area is the **flux density** and this is measured in Teslas.

32 Which image shows the correct direction of magnetic flux flow for the current carrying conductor?

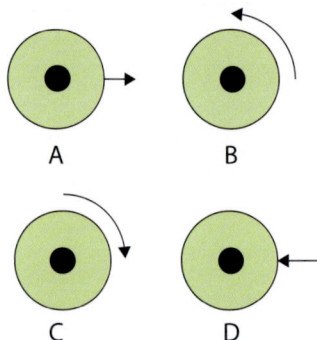

Correct answer b
Each of the conductors show a current flow towards rather than away. If the conductors showed a cross, this would indicate current flowing away.

Where current is flowing towards then flux would rotate around the conductor in an anti-clockwise direction.

Tip: Remember the screw rule!

EXAM PRACTICE 2

EXAM PRACTICE 2 QUESTIONS WITH ANSWERS: UNIT 003 ELECTRICAL SCIENTIFIC PRINCIPLES AND TECHNOLOGY

33 On the diagram, at which point of rotation would the maximum emf be induced by the simple alternator shown in the image?

Notes

- ○ a. A
- ○ b. B
- ● c. C
- ○ d. D

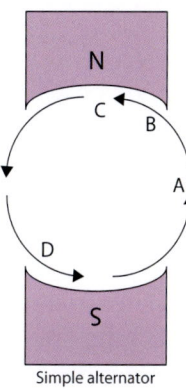

Simple alternator

Correct answer c

When a conductor is rotated in a magnetic field, the maximum emf would be induced when the conductor is at the strongest part of the field where the flux density is strongest. This would be at point C in the image.

34 What is the periodic time for a sinewave indicating a frequency of 50 Hz?

- ○ a. 5 ms
- ● b. 20 ms
- ○ c. 30 ms
- ○ d. 50 ms

Correct answer b

The frequency, measured in Hertz is the amount of cycles per second (1000 ms).

The periodic time for a sinewave is the time taken to complete one cycle so, if 50 cycles occur in 1000 milliseconds then one cycle occurs in

$$\frac{1000}{50} = 20 \; ms$$

Notes

35 What would be the value of emf induced from a simple alternator, having a 0.1 m conductor, rotating at a velocity of 40 metres per second within a magnetic field having a flux density of 4 Tesla?

- ◉ a. 16 V
- ○ b. 50 V
- ○ c. 100 V
- ○ d. 230 V

Correct answer a

The emf induced by an alternator is calculated using the Flux density (B), conductor length (L) and the velocity of rotation (v), so

$$e = BLv$$

So

$$e = 4 \times 0.1 \times 40 = 16\ V$$

36 Which quantity, indicated on the sinewave in the image, would be multiplied by 0.707 to obtain the rms voltage?

- ○ a. A
- ○ b. B
- ◉ c. C
- ○ d. D

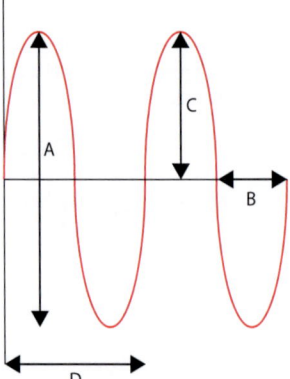

Correct answer c

The rms voltage is the peak voltage (indicated by C) multiplied by $\sqrt{3}$ or 0.707.

The other values indicated are:
A – Peak to peak value
B – Half cycle time
C – Periodic time

EXAM PRACTICE 2

EXAM PRACTICE 2 QUESTIONS WITH ANSWERS: UNIT 003 ELECTRICAL SCIENTIFIC PRINCIPLES AND TECHNOLOGY

37 What single electronic component provides half wave rectification?

- a. Capacitor.
- b. Inductor.
- ● c. Diode.
- d. Transistor.

Correct answer c
A diode is used to provide half wave rectification. Where full rectification is required, four diodes would be connected in a bridge.

38 What electronic component would be used to measure the water temperature within a central heating boiler?

- ● a. Thermistor.
- b. Invertor.
- c. Resistor.
- d. Transistor.

Correct answer a
A thermistor is a thermal resistor where the resistance changes proportional to the temperature it is exposed to.

39 What electronic component stores a charge?

- a. Inductor.
- b. Transistor.
- ● c. Capacitor.
- d. Resistor.

Correct answer c
Capacitors store charge whilst connected to a supply and would then release that charge if a load or resistor is connected across it.

40 What electronic component is commonly used as a light source or indicator?

- a. SCR
- b. PCP
- ● c. LED
- d. LDD

Correct answer c
An LED, or light emitting diode is now a very common component used in modern lighting systems. It is also very commonly used as an indicator light in electrical and electronic equipment.

Notes

Notes

Exam practice 2 answer key: Unit 003 electrical scientific principles and technologies

1 c	9 a	17 b	25 d	33 c
2 d	10 d	18 d	26 c	34 b
3 b	11 a	19 d	27 c	35 a
4 a	12 c	20 b	28 a	36 c
5 d	13 d	21 c	29 b	37 c
6 c	14 a	22 b	30 d	38 a
7 b	15 b	23 c	31 d	39 c
8 a	16 b	24 a	32 b	40 c

Exam Practice 3

Exam Practice 3	
Exam practice 3: Unit 103 (written paper) electrical scientific principles and technologies	78
Exam practice 3 questions with model answers: Unit 103 written paper	82

Exam practice 3: Unit 103 (written paper) electrical scientific principles and technologies

1. Show, using a block diagram, how wood, when used as an energy source, is used to produce electricity. (3 marks)

2. a) State **two** distribution voltages used in the UK other than 400/230 V. (2 marks)
   b) State **one** transmission voltage used in the UK. (1 mark)

3. Identify three renewable sources of energy which involve water power without the need for heat. (3 marks)

4. Describe how an 11 kV three-line supply is transformed into a three-phase four-wire 400 V distribution system. (3 marks)

5. Describe, with the aid of a diagram, how a core-type transformer is arranged. (3 marks)

6. a) State **one** loss that can occur in the winding of a transformer. (1 mark)
   b) State **one** loss that can occur in the core of a transformer. (1 mark)
   c) State **one** method of reducing core loss in a transformer. (1 mark)

7. A coil having both resistance and inductive reactance is connected to an AC supply. Draw an impedance triangle showing the relationship between these properties. (3 marks)

8. A coil with an inductance of 0.4 H and resistance 50 Ω is connected to a 230 V 50 Hz supply.

   Determine, showing all calculations, the impedance of the circuit. (3 marks)

EXAM PRACTICE 3

EXAM PRACTICE 3: UNIT 103 (WRITTEN PAPER) ELECTRICAL SCIENTIFIC PRINCIPLES AND TECHNOLOGIES

9 Describe, with the aid of a power triangle, the relationship between kW and kVA. (3 marks)

10 Determine, showing all calculations, the power factor of a 3 kW motor that draws 20 A from a 230 V single-phase supply. (3 marks)

11 a) State **two** examples of loads where power factor correction may be required. (2 marks)
   b) State what the cos ø relates to with regard to power factor. (1 mark)

12 A luminaire has a power factor of 0.4 lagging.

   a) State the component that can be installed to improve power factor. (1 mark)
   b) State how the component is connected to the luminaire. (1 mark)
   c) State the effect on the supply current once the component is installed. (1 mark)

13 Determine the neutral current using the values shown in the following image. (3 marks)

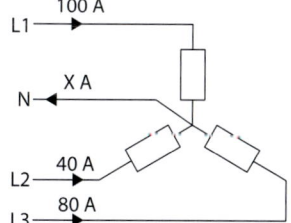

14 A three-phase balanced delta connected load is supplied at a line voltage of 400 V and draws a line current of 8 A.

   a) Determine the phase voltage. (1 mark)
   b) Determine the line current. (2 marks)

15 Draw a circuit diagram for a series wound DC machine identifying the key components. (3 marks)

Notes

Notes

16. a) Explain how the start winding creates starting torque in a capacitive start single-phase AC induction motor. (2 marks)
    b) State the component that switches out the start winding once the motor has started. (1 mark)

17. a) State **two** advantages for AC induction motors compared to wound rotor motors. (2 marks)
    b) Describe how the direction of a three-phase induction motor can be reversed. (1 mark)

18. Describe why a direct-on-line starter is used to control a smallpower machine. (3 marks)

19. State the internal component that causes an RCBO to disconnect under **each** of the following conditions.
    a) Residual current flows to earth. (1 mark)
    b) The circuit is loaded much higher than the nominal rating. (1 mark)
    c) A short circuit between Line and Neutral. (1 mark)

20. Describe, with the aid of a diagram, how a relay and two lamps are used to indicate that equipment is either energised or not. (3 marks)

21. Explain why cartridge fuses have sand around the elements. (3 marks)

22. Calculate the illuminance on a surface 6 m below a lamp that emits 4000 candelas. (3 marks)

23. Explain the purpose of the choke/ballast unit in a low-pressure fluorescent luminaire. (3 marks)

EXAM PRACTICE 3: UNIT 103 (WRITTEN PAPER) ELECTRICAL SCIENTIFIC PRINCIPLES AND TECHNOLOGIES

24 a) State how an incandescent lamp operates. (2 marks)
    b) What type of discharge luminaire is best suited to illuminating a large sports field with white light. (1 mark)

25 State the method of heat transfer used for **each** of the following heaters.
    a) Immersion water heater. (1 mark)
    b) Infrared space heater. (1 mark)
    c) Panel type wall heater. (1 mark)

26 Explain how the convection cycle works when heating a small room. (3 marks)

Notes

Exam practice 3 questions with model answers: Unit 103 written paper

1. Show, using a block diagram, how wood, when used as an energy source, is used to produce electricity. (3 marks)

   Model answer:
   The block diagram should indicate, in very simple terms, the process involved in producing electricity where the heat from the gas produces steam from water which, at high pressure, rotates a turbine which in turn, rotates an alternator.

   Marks are typically awarded for each stage in the process following the burning of the fuel.

   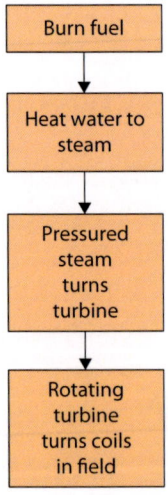

2. a) State two distribution voltages used in the UK other than 400/230 V. (2 marks)
   b) State one transmission voltage used in the UK. (1 mark)

   Model answer:
   a) Distribution voltages other than 400/230 V are 11 kV, 33 kV, 66 kV and 132 kV.
   b) Transmission voltages are 400 kV and 275 kV and 132 kV.

   Tip: Marks are awarded for the first answers given so there is no point in writing as many voltages down as you can remember!

EXAM PRACTICE 3

**EXAM PRACTICE 3 QUESTIONS WITH MODEL ANSWERS:
UNIT 103 WRITTEN PAPER**

3 Identify three renewable sources of energy which involve water power without the need for heat. (3 marks)

Model answer:
The three acceptable answers are

- Hydro
- Tidal
- Wave

Although burning fossil fuels can heat water, these involve heat and are not classed as renewable so would not be acceptable answers.

4 Describe how an 11 kV three-line supply is transformed into a three-phase four-wire 400 V distribution system. (3 marks)

Model answer:
The answer to this question could be presented using bullet points in order to ensure the key detail isn't lost in a large paragraph so

- 11 kV primary side of transformer wound in Delta
- 400 V secondary side of transformer wound in Star
- Neutral taken from Star point

If you prefer, a diagram could be used to show this, even though the question does not require one.

Marks are typically awarded for the three key stages indicated by the bullet points above.

Notes

Notes

5 Describe, with the aid of a diagram, how a core-type transformer is arranged. (3 marks)

Model answer:
The diagram should show a core type of core (not shell type) with the primary and secondary windings on each side of the core.

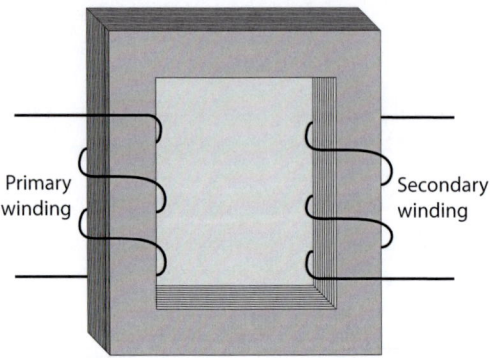

Marks are typically awarded for showing the

- Correct core
- Primary winding or input
- Secondary winding or output

The diagram **does not** need to be detailed showing laminations or insulators etc.

6 a) State **one** loss that can occur in the winding of a transformer. (1 mark)

b) State **one** loss that can occur in the core of a transformer. (1 mark)

c) State **one** method of reducing core loss in a transformer. (1 mark)

Model answer:
a) Copper loss.
b) Iron loss (or) hysteresis (or) eddy current loss (any one from these)
c) Laminating the core

Tip: where one or two words will answer the question, don't use any more as the keyword becomes buried and the answer may be missed.

7. A coil having both resistance and inductive reactance is connected to an AC supply. Draw an impedance triangle showing the relationship between these properties. (3 marks)

Model answer:
A suitable impedance triangle needs to be drawn indicating the three main properties.

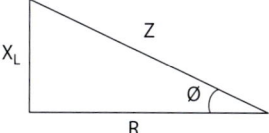

A suitable impedance triangle needs to be drawn indicating the three main properties

Marks are typically awarded for correctly indicating the:

- Resistance (R)
- Inductive reactance (X_L)
- Impedance (Z)

8. A coil of inductance 0.4 H and resistance 50 Ω is connected to a 230 V 50 Hz supply.

   Determine, showing all calculations, the impedance of the circuit. (3 marks)

Model answer:

Tip for all questions using a formula:
You must show your calculations as marks are awarded for them. You do not need to show the formula without values as shown below, any model answer in this book will show the full formula so you can see the values that should be used and how it is calculated.

$$X_L = 2\pi fL \text{ so } 2\pi \times 50 \times 0.4 = 125.7 \text{ Ω}$$

$$Z = \sqrt{X_L^2 + R^2} \text{ so}$$

$$\sqrt{125.7^2 + 50^2} = 135.3 \text{ Ω}$$

Marks are typically awarded for the formula used to obtain X_L, the formula to determine Z and the correct answer. If you make mistakes on the way through the process, you may still get two marks for using the formula correctly.

Notes

9 Describe, with the aid of a power triangle, the relationship between kW and kVA. (3 marks)

Model answer:
The diagram should include the base line as kW and the hypotenuse as kVA.

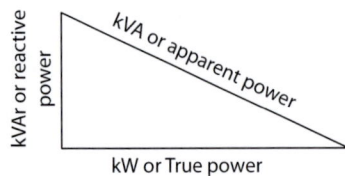

The relationship between the two values is the kVAr (power factor cos ø is a suitable alternative).
Marks are typically awarded for

- Correct labelling of kW
- Correct labelling of kVA
- Showing either kVAr (or Power factor)

10 Determine, showing all calculations, the power factor of a 3 kW motor that draws 20 A from a 230 V single-phase supply. (3 marks)

Model answer:

$$pf = \frac{kW}{kVA} \text{ or } \frac{W}{VA} \text{ so } \frac{3000}{230 \times 20} \, 0.65$$

Marks are typically awarded for

- Correctly used formula
- Conversion of kW into W or VA into kVA
- Calculation of the correct answer

11 a) State two examples of loads where power factor correction may be required. (2 marks)
    b) State what the cos ø relates to with regard to power factor. (1 mark)

Model answer:
a) Could include two of the following:
   - Discharge luminaires (lights)
   - Motors
   - Other suitable inductive loads involving reactance will also be accepted.

b) The power factor is the cosine of the angle by which the current leads or lags the voltage.

Marks are typically awarded for

a) Two good examples
b) Making a statement which includes cosine, angle, lead or lag

12 A luminaire has a power factor of 0.4 lagging.

   a) State the component that can be installed to improve power factor. (1 mark)
   b) State how the component is connected to the luminaire. (1 mark)
   c) State the effect on the supply current once the component is installed. (1 mark)

Model answer:

a) Capacitor
b) In parallel to the supply terminals
c) Reduce

Marks are typically awarded in b) for the word parallel but the context of what it is in parallel to is required.

13 Determine the neutral current using the values shown in the image below. (3 marks)

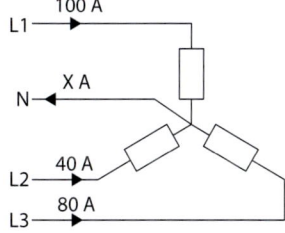

Model answer:
The neutral current may be determined using any of the following three methods

- Triangle
- Calculation
- Phasor

The method shown below is the calculated method but graphical methods are acceptable. Answers within ±5 A will be accepted.

$$\sqrt{100^2 + 40^2 + 80^2 - (100 \times 40) - (100 \times 80) - (40 \times 80)} = 53 \text{ A}$$

Marks are typically awarded for

- Method presentation
- Accuracy
- Calculating a correct answer

14 A three-phase balanced delta connected load is supplied at a line voltage of 400 V and draws a line current of 8 A.

   a) Determine the phase voltage. (1 mark)
   b) Determine the line current. (2 marks)

Model answer:

a) $V_L = V_P$ then $V_P = 400$ V

b) $I_P = \dfrac{I_L}{\sqrt{3}} = \dfrac{8}{\sqrt{3}} = 4.61$ A

Marks are typically awarded for

- 400 V
- I_P formula
- I_P correct answer

15 Draw a circuit diagram illustrating a series wound DC machine identifying the key components. (3 marks)

Model answer:

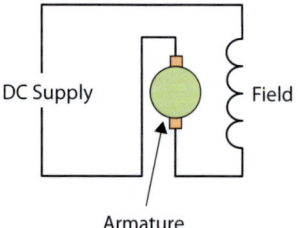

Drawings must show the field and armature in series.

Marks are typically awarded for

- Correct diagram (1 mark)
- Labelled field (1 mark)
- Labelled armature (1 mark)

EXAM PRACTICE 3

**EXAM PRACTICE 3 QUESTIONS WITH MODEL ANSWERS:
UNIT 103 WRITTEN PAPER**

16 a) Explain how the start winding creates starting torque in a capacitive start single-phase AC induction motor. (2 marks)
   b) State the component that switches out the start winding once the motor has started. (1 mark)

Model answer:
a) The capacitor creates a phase shift which induces a second magnetic field.
b) Centrifugal switch

Marks are typically awarded for

a)
- Phase shift or similar key wording
- Magnetic field

b)
- Centrifugal switch

17 a) State **two** advantages for AC induction motors compared to wound rotor motors. (2 marks)
   b) Describe how the direction of a three-phase induction motor can be reversed. (1 mark)

Model answer:
a) Answers may include
   - Low cost maintenance (no brushes or slip rings to maintain)
   - Low cost purchase or cheaper to manufacturer (similar wording acceptable)
b) Swap any two phases of the supply.

Marks are typically awarded for

a) Any two from:
   - low cost
   - cheap to buy
   - less parts
   - easy maintenance
   - less parts
   - less to go wrong
   - other suitable
b) swapping phases

18 Describe why a direct-on-line starter is used to control a smallpower machine. (3 marks)

Model answer:
Marks are typically awarded for the first three reasons, which should include:

- starter provides overload protection
- provides an emergency stop facility
- remote emergency stop buttons may be wired in
- provides undervoltage protection (preventing the motor automatically starting after a power failure)
- remote start facility may be wired in
- other suitable

Tip: Duplications will not be awarded marks. For example

- **additional stop buttons can be fitted**
- **stop buttons at various locations can be used**
- **provides overload protection**

The above will only give two marks as the first two are essentially the same answers.

19 State the internal component that causes an RCBO to disconnect under each of the following conditions.

   a) Residual current flows to earth. (1 mark)
   b) The circuit is loaded much higher than the nominal rating. (1 mark)
   c) A short circuit between Line and Neutral. (1 mark)

Model answer:

a) RCD element
b) Bi-metallic or overload mechanism
c) Magnetic tripping mechanism

Answers are typically awarded for answers that are describing the same component parts as each of the above.

EXAM PRACTICE 3
**EXAM PRACTICE 3 QUESTIONS WITH MODEL ANSWERS:
UNIT 103 WRITTEN PAPER**

20 Describe, with the aid of a diagram, how a relay and two lamps are used to indicate that equipment is energised or not. (3 marks)

Model answer:

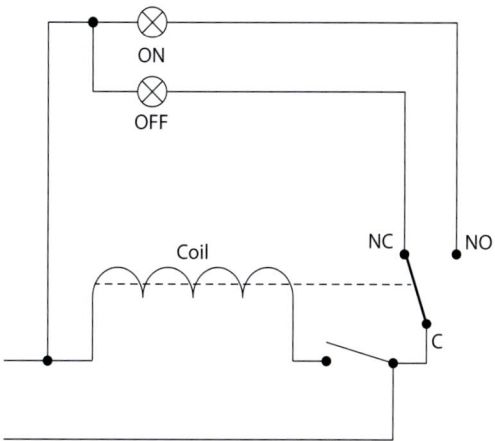

Note that the equipment would be wired in parallel to the coil.

Marks are typically awarded for

- Relay NO contacts to on lamp
- Relay NC contacts to off lamp
- General circuit layout

21 Explain why cartridge fuses have sand around the elements. (3 marks)

Model answer:
The sand gives the fuse a high breaking capacity as the sand turns to glass under high current fault conditions absorbing the blast energy.

Marks are typically awarded for keyword phrases in the description which are similar to

- Sand turns to glass
- High fault currents
- Absorbing energy

Notes

22 Calculate the illuminance on a surface 6 m below a lamp that emits 4000 candelas. (3 marks)

Model answer:
Using the inverse square law

$$E = \frac{I}{d^2} \quad \text{so} \quad \frac{4000}{6^2} = 111.1 \text{ Lux}$$

Marks are typically awarded for

- Formula
- Correct value
- Correct unit (Lux)

23 Explain the purpose of the choke/ballast unit in a low-pressure fluorescent luminaire. (3 marks)

Model answer:

- Creates a discharge voltage on start-up
- Limits lamp current demand once started
- By mutual induction

Marks are typically awarded for a description which includes

- How the unit starts the lamp
- How he unit limits current
- That it works by mutual induction

24 a) State how an incandescent lamp operates. (2 marks)
   b) What type of discharge luminaire is best suited to illuminating a large sports field with white light. (1 mark)

Model answer:
a) Answer to include (using key terms)
   - Resistive filament
   - Glows white hot
b) Metal halide (LED is not a discharge luminaire)

EXAM PRACTICE 3

**EXAM PRACTICE 3 QUESTIONS WITH MODEL ANSWERS:
UNIT 103 WRITTEN PAPER**

25 State the method of heat transfer used for each of the following heaters.

    a) Immersion water heater. (1 mark)
    b) Infra-red space heater. (1 mark)
    c) Panel type wall heater. (1 mark)

Model answer:

a) Conduction
b) Radiant
c) Convection

Marks are typically awarded for the first answer given in each case.

26 Explain how the convection cycle works when heating a small room. (3 marks)

Model answer:

- Air heated by heater rises
- Air cools and falls
- Cool air then enters bottom of heater and heated

Marks are typically awarded for the three main parts of the convection cycle such as

- Warmed air rises
- Air then cools and falls

Warm air leaving top of heater is replaced by lower cool air.

Notes

Notes

EXAM PRACTICE 4

Exam Practice 4

Exam practice 4: Unit 004 understand design and installation practices and procedures   96

Exam practice 4 questions with model answers: Unit 004 understand design and installation practices and procedures   104

Exam practice 4 answer key: Unit 004 understand design and installation practices and procedures   118

Exam practice 4: Unit 004 understand design and installation practices and procedures

1. Which document is non-statutory?

   - a. Electricity at Work Regulations.
   - b. BS 7671: Requirement for Electrical Installations.
   - c. Health & Safety at Work etc. Act.
   - d. Electricity Safety, Quality & Continuity Regulations.

2. Which document will **most** likely contain instructions for connecting a particular item of electrical equipment?

   - a. BS 7671: IET Wiring Regulations.
   - b. IET GN1: *Selection and Erection*.
   - c. Manufacturer's instructions.
   - d. Electricity at Work Regulations.

3. Before work begins, an electrician notices damage to a ceiling in a room which is to be worked in. What is the correct action to take?

   - a. None as the client must be aware of it.
   - b. Notify the client so you can't be blamed.
   - c. Repair it before work starts near it.
   - d. Ensure the client repairs it before you start.

4. What is an essential safety consideration when planning and setting up a construction site?

   - a. Location of the canteen.
   - b. Location for visitor parking.
   - c. Identify local transport links.
   - d. Identify evacuation routes.

EXAM PRACTICE 4

EXAM PRACTICE 4: UNIT 004 UNDERSTAND DESIGN AND INSTALLATION PRACTICES AND PROCEDURES

5. How can a client's laminate flooring be **best** protected when it must be crossed by site operatives, plant and machinery when accessing the site?

    - a. Cover the floor with hardboard.
    - b. Cover the floor with a dustsheet.
    - c. Cover the floor with plastic sheeting.
    - d. Cover the floor with newspaper.

6. What is the part of the cable marked B?

    - a. Insulator.
    - b. Conductor.
    - c. Sheath.
    - d. Armour.

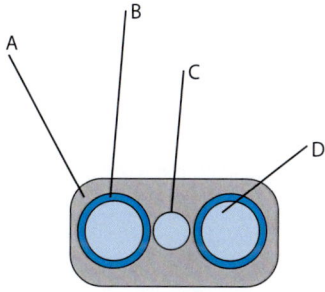

7. What material is used for the core of a fibre-optic cable?

    - a. Copper.
    - b. Steel.
    - c. Glass.
    - d. Brass.

8. What insulation is used in MICC cables?

    - a. PVC.
    - b. XLPE.
    - c. Thermoplastic.
    - d. Magnesium Oxide.

9 What type of conduit fitting is shown in the following image?

○ a. Tangent angle.
○ b. Cosine angle.
○ c. Straight angle.
○ d. Opposing angle.

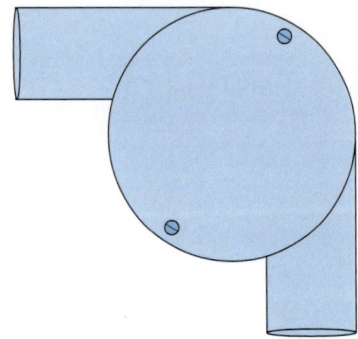

10 What is the cable support system shown in the following image?

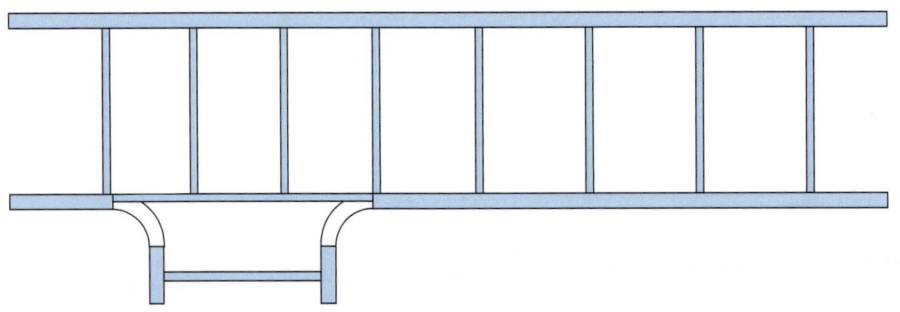

○ a. Ladder.
○ b. Tray.
○ c. Basket.
○ d. Track.

11 What type of wiring system is fixed using saddles?

○ a. Trunking.
○ b. Basket.
○ c. Conduit.
○ d. Tray.

EXAM PRACTICE 4

EXAM PRACTICE 4: UNIT 004 UNDERSTAND DESIGN AND INSTALLATION PRACTICES AND PROCEDURES

12. What is the **minimum** suitable size of trunking, in mm^2, for the following PVC stranded cables?

    - 16 x 2.5 mm^2
    - 18 x 4 mm^2
    - 6 x 10 mm^2
    - 8 x 16 mm^2

    - a. 50 x 38
    - b. 50 x 50
    - c. 75 x 25
    - d. 75 x 38

13. How many 1.5 mm^2 stranded conductors are permitted within a 1 m straight run of 20 mm conduit?

    - a. 12
    - b. 13
    - c. 14
    - d. 15

14. What conduit becomes brittle when installed in direct sunlight?

    - a. Black PVC.
    - b. White PVC.
    - c. Black enamelled.
    - d. Galvanised.

15. What is a masonry drill bit suitable for drilling?

    - a. Wood.
    - b. Plastic.
    - c. Glass.
    - d. Brick.

16. What is the purpose of a chalk line?

    - a. To mark off lengths of tray for cutting.
    - b. To give a straight line on a wall for fixings.
    - c. To write temporary instructions on a floor.
    - d. To wrap around conduit to give a straight cut.

Notes

17 What would be used to cut a 20 mm diameter round opening in a metal trunking?

- a. Hole-saw.
- b. Auger-bit.
- c. Hacksaw.
- d. Masonry bit.

18 What is the **main** consideration when selecting a fixing device for a large SWA cable, installed above a suspended ceiling, in an office location?

- a. Corrosion.
- b. Aesthetics.
- c. Load-bearing.
- d. Finish.

19 What type of device is shown in the following image?

- a. Crampet.
- b. Spacer-bar saddle.
- c. Distance saddle.
- d. Cleat.

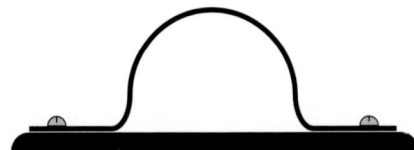

20 What is the **maximum** recommended distance between supports for a horizontal 25 mm metal conduit?

- a. 1.75 m
- b. 2.00 m
- c. 2.25 m
- d. 2.50 m

EXAM PRACTICE 4: UNIT 004 UNDERSTAND DESIGN AND INSTALLATION PRACTICES AND PROCEDURES

21 What, from the following methods, must be used to mechanically protect a cable, which is concealed in a wall, 15 mm deep, and is run as shown in the image?

- a. RCD protection only.
- b. Unearthed capping.
- c. Plastic oval conduit.
- d. Earthed metal conduit.

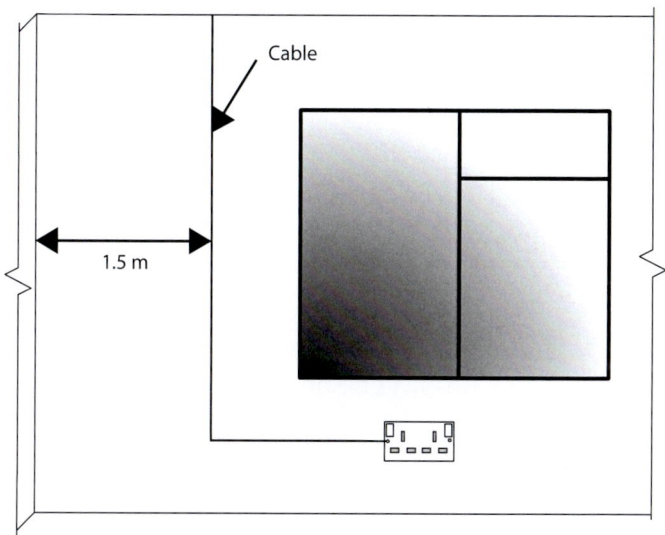

22 What is the **most** professional method of dealing with damaged walls through chasing for cable runs?

- a. Leave the rubble and walls for the client to sort out.
- b. Include the cost of a professional plasterer.
- c. Put all rubble in the bin and wallpaper over all the chases.
- d. Fill all chases with a resin based, waterproof sealant.

23 What is a fault between two lines of a three-phase circuit defined as?

- a. Earth fault.
- b. Overload.
- c. Open circuit.
- d. Short circuit.

Notes

Notes

24 What is the **maximum** permitted earth fault loop impedance for a final circuit protected by a 20 A BS 88-2 fuse, as given in the IET On-site Guide?

- a. 2.21 Ω
- b. 2.03 Ω
- c. 1.46 Ω
- d. 1.30 Ω

25 What is the **maximum** permitted earth fault loop impedance for a final circuit protected by a 32 A Type C circuit breaker to BS EN 60898, as given in the IET On-site Guide?

- a. 0.55 Ω
- b. 0.79 Ω
- c. 1.10 Ω
- d. 1.38 Ω

26 What is the **maximum** value of fault current that can be safely disconnected by a BS 88-3 type I device?

- a. 6 kA
- b. 10 kA
- c. 16 kA
- d. 80 kA

27 What should be achieved by correct selectivity under fault conditions?

- a. The protective device at the origin should disconnect all faults.
- b. The protective device on the supply side nearest the fault should disconnect.
- c. The protective device within the DNO's service head should disconnect all faults.
- d. The protective device on the load side nearest the fault should disconnect.

EXAM PRACTICE 4
EXAM PRACTICE 4: UNIT 004 UNDERSTAND DESIGN AND INSTALLATION PRACTICES AND PROCEDURES

28 A final radial circuit is to be installed, forming part of a TN-S earthing arrangement, where the circuit is protected by a 32 A/30 mA Type B RCBO and the circuit is wired using 4/1.5 mm² flat profile twin and CPC cable.

What is the **maximum** length for this final circuit as given in the IET On-site Guide?

- a. 33 m.
- b. 49 m.
- c. 81 m.
- d. 104 m.

29 What is a circuit linking together two distribution boards known as?

- a. Final circuit.
- b. Closed circuit.
- c. Distribution circuit.
- d. Ring-final circuit.

30 What is the purpose of a contactor, within a Direct Online (DOL) starter, controlling a machine?

- a. To enable manual operation of the machine in the event of a power failure.
- b. To prevent automatic restarting of the machine following a power failure.
- c. To enable the machine and final circuit to have RCD protection.
- d. To prevent an RCD in the final circuit operating on start-up.

Notes

Notes

Exam practice 4 questions with model answers: Unit 004 understand design and installation practices and procedures

1. Which document is non-statutory?

   - a. Electricity at Work Regulations.
   - **b. BS 7671: Requirement for Electrical Installations.**
   - c. Health & Safety at Work etc. Act.
   - d. Electricity Safety, Quality & Continuity Regulations.

   Correct answer b
   All the other documents listed are statutory.

   BS 7671 are non-statutory regulations, they may however, be used in a court of law to claim compliance with statutory regulations.

2. Which document will **most** likely contain instructions for connecting a particular item of electrical equipment?

   - a. BS 7671: IET Wiring Regulations.
   - b. IET GN1: *Selection and Erection*.
   - **c. Manufacturer's instructions.**
   - d. Electricity at Work Regulations.

   Correct answer c
   Manufacturer's instructions will contain instructions or diagrams for connecting a particular item of equipment.

3. Before work begins, an electrician notices damage to a ceiling in a room which is to be worked in. What is the correct action to take?

   - a. None as the client must be aware of it.
   - **b. Notify the client so you can't be blamed.**
   - c. Repair it before work starts near it.
   - d. Ensure the client repairs it before you start.

   Correct answer b
   Always carry out a pre-work survey to determine any pre-existing damage. If it is left to the end, you may be blamed and requested to pay for repairs. This isn't good for relationships.

EXAM PRACTICE 4

EXAM PRACTICE 4 QUESTIONS WITH MODEL ANSWERS: UNIT 004 UNDERSTAND DESIGN AND INSTALLATION PRACTICES AND PROCEDURES

4 What is an essential safety consideration when planning and setting up a construction site?

- ○ a. Location of the canteen.
- ○ b. Location for visitor parking.
- ○ c. Identify local transport links.
- ● d. Identify evacuation routes.

Correct answer d
Safety considerations are an essential part of planning a worksite and this must include safe evacuation and assembly points in the event of an emergency such as a fire.

5 How can a client's laminate flooring be **best** protected when it must be crossed by site operatives, plant and machinery when accessing the site?

- ● a. Cover the floor with hardboard.
- ○ b. Cover the floor with a dustsheet.
- ○ c. Cover the floor with plastic sheeting.
- ○ d. Cover the floor with newspaper.

Correct answer a
As the floor is to be crossed by many people and potentially heavy objects, hardboard will protect the floor from and scratches or dents as well as dirt.

6 What is the part of the cable marked B?

- ○ a. Insulator.
- ● b. Conductor.
- ○ c. Sheath.
- ○ d. Armour.

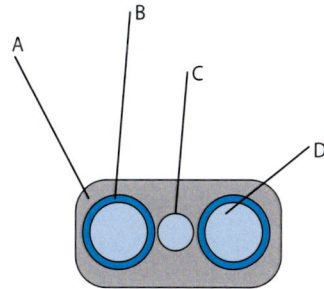

Correct answer b
b is indicating one of the cable cores which acts as a conductor.

Notes

7 What material is used for the core of a fibre-optic cable?

  ○ a. Copper.
  ○ b. Steel.
  ● c. Glass.
  ○ d. Brass.

  Correct answer c
  As fibre optic cables are used to transmit information using light, the cable core is made of thin glass.

8 What insulation is used in MICC cables?

  ○ a. PVC.
  ○ b. XLPE.
  ○ c. Thermoplastic.
  ● d. Magnesium Oxide.

  Correct answer d
  The insulation between the cores is magnesium oxide which as a white powder like substance.

9 What type of conduit fitting is shown in the following image?

  ● a. Tangent angle.
  ○ b. Cosine angle.
  ○ c. Straight angle.
  ○ d. Opposing angle.

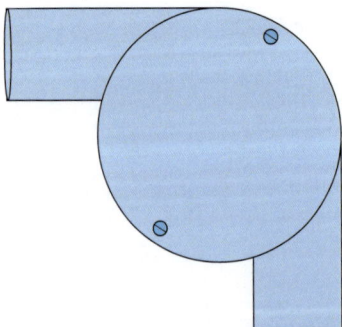

  Correct answer a
  The image shows a tangent angle box. When the spouts flow into the curve of the fitting, it is called a tangent.

EXAM PRACTICE 4

EXAM PRACTICE 4 QUESTIONS WITH MODEL ANSWERS: UNIT 004 UNDERSTAND DESIGN AND INSTALLATION PRACTICES AND PROCEDURES

10 What is the cable support system shown in the following image?

Notes

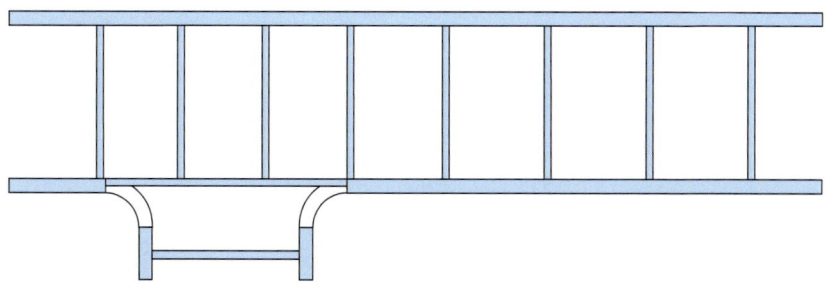

- ● a. Ladder.
- ○ b. Tray.
- ○ c. Basket.
- ○ d. Track.

Correct answer a
The image shows a cable ladder system which is generally used for heavy cables such as SWA cables.

11 What type of wiring system is fixed using saddles?

- ○ a. Trunking.
- ○ b. Basket.
- ● c. Conduit.
- ○ d. Tray.

Correct answer c
Conduit is fitted to a wall using a saddle such as a spacer bar saddle.

12 What is the **minimum** suitable size of trunking, in mm², for the following PVC stranded cables?

- 16 x 2.5 mm²
- 18 x 4 mm²
- 6 x 10 mm²
- 8 x 16 mm²

- ○ a. 50 x 38
- ○ b. 50 x 50
- ○ c. 75 x 25
- ● d. 75 x 38

Correct answer d
Using table E5 in the IET On-site Guide, determine the total PVC stranded cable factors-

Notes

From On-Site Guide BS 7671:2008+A3:2015 – Extract from Table E5 Cable factors for trunking p143

Type of conductor	Conductor cross-sectional area (mm²)	PVC BS 6004 Cable factor
Stranded	2.5	12.6
Stranded	4	16.6
Stranded	10	35.3
Stranded	16	47.8

$$2.5 \text{ mm}^2 = 12.6 \times 16 = 201.6$$

$$4.0 \text{ mm}^2 = 16.6 \times 18 = 298.8$$

$$10 \text{ mm}^2 = 35.3 \times 6 = 211.8$$

$$16 \text{ mm}^2 = 47.8 \times 8 = 382.4$$

Total = 1094.6

Using table E6 from IET On-site Guide:

From On-Site Guide BS 7671:2008+A3:2015 – Extract from Table E6 Factors for trunking p144

Dimensions of trunking (mm x mm)	Factor
75 x 38	1146

The minimum size trunking having a factor larger than 1094.6 is a 75 x 38 with a factor of 1146.

---

13 How many 1.5 mm² stranded conductors are permitted within a 1 m straight run of 20 mm conduit?

○ a. 12
○ b. 13
● c. 14
○ d. 15

Correct answer c
Using table E1 of the IET On-site Guide, a factor for 1.5 mm² stranded cable is 31.

EXAM PRACTICE 4

EXAM PRACTICE 4 QUESTIONS WITH MODEL ANSWERS: UNIT 004 UNDERSTAND DESIGN AND INSTALLATION PRACTICES AND PROCEDURES

Type of conductor	Conductor cross-sectional area (mm²)	Cable factor
Stranded	1.5	31

Using table E2 in the IET On-site Guide, a 20 mm conduit has a factor of 460.

Conduit diameter (mm)	Conduit factor (mm)
20	460

So

$$460 \div 31 = 14.8$$

So the minimum whole cables is 14.

Notes

From On-Site Guide BS 7671:2008+A3:2015 – Extract from Table E1 Cable factors for use in conduit in short straight runs p 140

From On-Site Guide BS 7671:2008+A3:2015 – Extract from Table E2 Conduit factors for use in short straight runs p140

14 What conduit becomes brittle when installed in direct sunlight?

- a. Black PVC.
- ● b. White PVC.
- c. Black enamelled.
- d. Galvanised.

Correct answer b
White PVC conduit is unsuitable in direct sunlight as the conduit becomes brittle and breaks or cracks.

15 What is a masonry drill bit suitable for drilling?

- a. Wood.
- b. Plastic.
- c. Glass.
- ● d. Brick.

Correct answer d
Masonry drill bits are specifically designed to drill into brick or concrete type.

Notes

16 What is the purpose of a chalk line?

- ○ a. To mark off lengths of tray for cutting.
- ● b. To give a straight line on a wall for fixings.
- ○ c. To write temporary instructions on a floor.
- ○ d. To wrap around conduit to give a straight cut.

Correct answer b
Chalk lines are stretched across a surface such as a wall then 'pinged' to leave a straight chalk line on the surface meaning the line itself can be removed. Chalk lines are very useful for marking long straight lines for fixings.

17 What would be used to cut a 20 mm diameter round opening in a metal trunking?

- ● a. Hole-saw.
- ○ b. Auger-bit.
- ○ c. Hacksaw.
- ○ d. Masonry bit.

Correct answer a
Hole-saws are used to cut round holes in material such as metal.

18 What is the main consideration when selecting a fixing device for a large SWA cable, installed above a suspended ceiling, in an office location?

- ○ a. Corrosion.
- ○ b. Aesthetics.
- ● c. Load-bearing.
- ○ d. Finish.

Correct answer c
If the fixing is indoors and behind a suspended ceiling above the heads of person, load-bearing is the biggest consideration.

EXAM PRACTICE 4

EXAM PRACTICE 4 QUESTIONS WITH MODEL ANSWERS: UNIT 004 UNDERSTAND DESIGN AND INSTALLATION PRACTICES AND PROCEDURES

Notes

19 What type of device is shown in the following image?

- ○ a. Crampet.
- ● b. Spacer-bar saddle.
- ○ c. Distance saddle.
- ○ d. Cleat.

Correct answer b
The spacer-bar saddle shown is used to fix conduit.

20 What is the **maximum** recommended distance between supports for a horizontal 25 mm metal conduit?

- ● a. 1.75 m
- ○ b. 2.00 m
- ○ c. 2.25 m
- ○ d. 2.50 m

Correct answer a
Using Table D3 in the IET On-site Guide, a horizontal 25 mm (16<d≤25) Rigid metal distance is shown as 1.75 m.

Nominal diameter of conduit, d (mm)	Maximum distance between supports
	Rigid metal
	Horizontal
1	
16 < d ≤ 40	1.75

From On-Site Guide BS 7671:2008+A3:2015 – Extract from Table D3 Spacings of supports for conduits p137

Notes

21 What, from the following methods, must be used to mechanically protect a cable, which is concealed in a wall, 15 mm deep, and is run as shown in the image?

○ a. RCD protection only.
○ b. Un-earthed capping.
○ c. Plastic oval conduit.
● d. Earthed metal conduit.

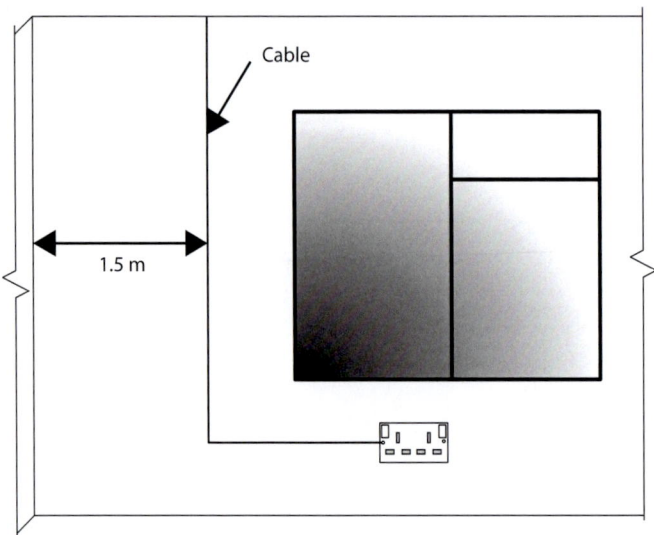

Answer d

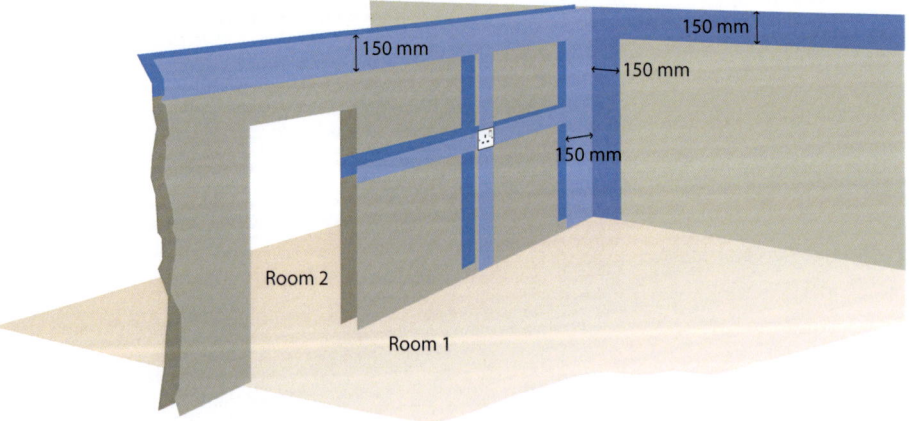

Cable concealed in a wall must fall into the zones of protection as described in Regulation 522.6.202, shown above, with no further mechanical protection. This example doesn't.

EXAM PRACTICE 4

EXAM PRACTICE 4 QUESTIONS WITH MODEL ANSWERS: UNIT 004 UNDERSTAND DESIGN AND INSTALLATION PRACTICES AND PROCEDURES

In this example, Regulation 522.6.204 states that cables concealed in a wall must be protected by a range of methods, one being earthed metallic conduit.

The above is also detailed in the IET On-site Guide Section 7.3.2

22 What is the **most** professional method of dealing with damaged walls through chasing for cable runs?

- a. Leave the rubble and walls for the client to sort out.
- ● b. Include the cost of a professional plasterer.
- c. Put all rubble in the bin and wallpaper over all the chases.
- d. Fill all chases with a resin based, waterproof sealant.

Correct answer b
If work involves major restoration, this should be carried out by professionals and included in the cost of the work.

23 What is a fault between two-lines of a three-phase circuit defined as?

- a. Earth fault.
- b. Overload.
- c. Open circuit.
- ● d. Short circuit.

Correct answer d
Where two **live** conductors are in contact in a circuit, due to a fault, the fault is classed as a short circuit.

24 What is the **maximum** permitted earth fault loop impedance for a final circuit protected by a 20 A BS 88-2 fuse, as given in the IET On-site Guide?

- a. 2.21 Ω
- b. 2.03 Ω
- c. 1.46 Ω
- ● d. 1.30 Ω

Correct answer d
A 20 A final circuit must disconnect within 0.4 seconds.

From Table B3(i) of the IET On-site Guide which gives maximum values for BS 88-2 devices for 0.4 s disconnection, the maximum value permitted, for all CPC sizes is 1.3 Ω

Notes

From On-Site Guide BS 7671:2008+A3:2015 – Extract from Table B3(i) 0.4 second disconnection (final circuits not exceeding 32 A in TN systems) p122

Protective conductor (mm²)	Fuse rating: 20 A
1.0	1.3
1.5	1.3
≥	1.3

25 What is the **maximum** permitted earth fault loop impedance for a final circuit protected by a 32 A Type C circuit breaker to BS EN 60898, as given in the IET On-site Guide?

- ● a. 0.55 Ω
- ○ b. 0.79 Ω
- ○ c. 1.10 Ω
- ○ d. 1.38 Ω

Correct answer a
Using Table B6 in the IET On-site Guide, a 32 A Type C device has a maximum permitted value of 0.55 Ω.

Circuit breaker type	Circuit breaker rating (amperes)
	32
3&C	0.55

From On-Site Guide BS 7671:2008+A3:2015 - Extract from Table B6 Circuit breakers.

Maximum measured earth fault loop impedance (in ohms) at ambient temperature where the overcurrent device is a circuit breaker to BS 3871 or BS EN 60898 or RCBO to BS EN 61009 p125.

26 What is the **maximum** value of fault current that can be safely disconnected by a BS 88-3 type I device?

- ○ a. 6 kA
- ○ b. 10 kA
- ● c. 16 kA
- ○ d. 80 kA

Correct answer c
Using Table 7.2.7(i) of the IET On-site Guide, the maximum rated short circuit capacity of a BS 88-3 type I device is 16 kA.

From On-Site Guide BS 7671:2008+A3:2015 – Extract from Table 7.2.7(i) Rated short-circuit capacities p74

Device type	Rated short circuit capacity (kA)
BS-88-3 type I	16
BS-88 type II	31.5

27 What should be achieved by correct selectivity under fault conditions?

- a. The protective device at the origin should disconnect all faults.
- **b. The protective device on the supply side nearest the fault should disconnect.** ●
- c. The protective device within the DNO's service head should disconnect all faults.
- d. The protective device on the load side nearest the fault should disconnect.

Correct answer b

Selectivity, which used to be known as discrimination, is careful co-ordination of protective devices in series with one another. Good selectivity means the device nearest the fault, on the supply side of the fault, should disconnect first minimising any inconvenience.

28 A final radial circuit is to be installed, forming part of a TN-S earthing arrangement, where the circuit is protected by a 32 A/30 mA Type B RCBO and the circuit is wired using 4/1.5 mm² flat profile twin and CPC cable?

What is the maximum length for this final circuit as given in the IET On-site Guide?

- **a. 33 m.** ●
- b. 49 m.
- c. 81 m.
- d. 104 m.

Correct answer a

Using Table 7.1(i) of the IET On-site Guide, a circuit protected by a 32 A Type B RCBO having 30 mA RCD protection, and forming part of a TN-S system, the maximum length is 33 m.

Notes

From On-Site Guide BS 7671: 2008+A3: 2015 – Extract from Table 7.1(i)

Protective device		Cable size mm²	Allowed installation method	$Z_s \leq 0.8\Omega$ TN-S		$Z_s \leq 0.35 \Omega$ TN-C-S	
Rating (A)	Type			RCD 30 mA	No RCD	RCD 30 mA	No RCD
32	cb/RCBO Type B	4.0 / 1.5	C	33	31zs	33	33
	cb/RCBO Type C		C	33	NPzs	33	14zs
	cb/RCBO Type D		C	33	NPzs	33	NPzs

Maximum cable length for a 230 V final circuit in domestic premises and similar using 70 °C thermoplastic (PVC) insulated and sheathed flat cable.

29 What is a circuit linking together two distribution boards known as?

- a. Final circuit.
- b. Closed circuit.
- ● c. Distribution circuit.
- d. Ring-final circuit.

Correct answer c
A distribution circuit is a circuit supplied from one distribution board and supplies another.

EXAM PRACTICE 4

EXAM PRACTICE 4 QUESTIONS WITH MODEL ANSWERS: UNIT 004 UNDERSTAND DESIGN AND INSTALLATION PRACTICES AND PROCEDURES

30 What is the purpose of a contactor, within a Direct Online (DOL) starter, controlling a machine?

- a. To enable manual operation of the machine in the event of a power failure.
- b. To prevent automatic restarting of the machine following a power failure.
- c. To enable the machine and final circuit to have RCD protection.
- d. To prevent an RCD in the final circuit operating on start-up.

Correct answer b

One of the main reasons for installing a DOL starter to a machine is so, in the event of a power failure, the contactor drops out meaning the machine cannot suddenly re-start when the power is restored. If the machine did suddenly restart after a power cut, there is a risk of injury to anyone near the machine.

Notes

Exam practice 4 answer key: Unit 004 understand design and installation practices and procedures

1 a	7 c	13 c	19 b	25 a
2 c	8 d	14 b	20 a	26 c
3 b	9 a	15 d	21 d	27 b
4 d	10 a	16 b	22 b	28 a
5 a	11 c	17 a	23 d	29 c
6 b	12 d	18 c	24 d	30 b

Exam Practice 5

Exam Practice 5

Exam practice 5: Unit 005 understand how to plan and oversee electrical work activities — 120

Exam practice 5 questions with model answers: Unit 005 understand how to plan and oversee electrical work activities — 124

Exam practice 5 answer key: Unit 005 understand how to plan and oversee electrical work activities — 130

Exam practice 5: Unit 005 understand how to plan and oversee electrical work activities

1. What is the **most** suitable method of contacting a colleague to remind them, without too much disturbance late at night, that you will be late arriving on site the next day?

   - a. Electronic mail.
   - b. Text message.
   - c. Postal letter.
   - d. Telephone call.

2. What is the **most** reliable method of communicating with other trades on a small site, to ensure all parties understand a minor problem that has occurred?

   - a. Face to face communication.
   - b. Electronic mail (email).
   - c. Social media messaging.
   - d. Written posted letter.

3. How should you initially communicate with a work colleague who is lacking motivation due to personal issues?

   - a. Informal conversation.
   - b. Sending a formal letter.
   - c. Email with final written warning.
   - d. Discussion via open social media.

4. What is the **most** appropriate method of ensuring many trades are up to date and understand major alterations to a contract?

   - a. Text message.
   - b. E-mail.
   - c. Site meeting.
   - d. Twitter.

EXAM PRACTICE 5

EXAM PRACTICE 5: UNIT 005 UNDERSTAND HOW TO PLAN AND OVERSEE ELECTRICAL WORK ACTIVITIES

5 What card can prove a person is fully qualified as an electrician?

- a. CSCS.
- b. ECS.
- c. PASMA.
- d. ID.

6 What is the **most** efficient method for selecting a shortlist of candidates, from many applications, for a job role?

- a. Hold interviews.
- b. Have trial periods.
- c. Request candidate CVs.
- d. Visit their current employers.

7 What needs to be considered when a work schedule is severely disrupted due to bad weather?

- a. Availability of suitable labour.
- b. Availability of delivered materials.
- c. Suitability of specified accessories.
- d. Suitability of installed equipment.

8 What is the purpose of a variation order?

- a. To inform the client of amendments to the current Wiring Regulations.
- b. To maintain a record of amendments to the original agreed contract.
- c. To allow persons on site to alter the contract to make work easier.
- d. To give a wholesaler permission to substitute items without discussion.

9 What document allows an organisation to check that delivered materials are in accordance to the original order?

- a. Variation order.
- b. Site diary.
- c. Delivery note.
- d. Take-off sheets.

Notes

Notes

10 What document would be used to determine the quantity of accessories needed for a contract?

- a. Site plans.
- b. Site diary.
- c. Delivery note.
- d. Critical network.

11 What, from the following, is identified by a Gantt chart?

- a. Labour requirements at various stages of the contract.
- b. Quantities of materials left over at the end of a contract.
- c. How much material was taken to landfill or recycled.
- d. The energy performance of an installed electrical system.

12 What is the **main** purpose of a material schedule?

- a. To ensure waste electrical materials are disposed of correctly.
- b. To ensure materials are stored on site at the correct temperature.
- c. To ensure that the correct type of materials are purchased.
- d. To ensure materials are handled in accordance with COSHH.

13 What document should be agreed upon by the client and contractor, before work begins, to ensure all accessories are fit for purpose?

- a. Variation order.
- b. Materials schedule.
- c. Day-work sheets.
- d. Delivery notes.

14 A client specifies that they want a particular socket-outlet outdoors which they do not realise is non-compliant. The correct type needed is much more expensive.

What is the **most** likely outcome if a contractor doesn't discuss this issue with the client before a contract is signed?

- a. The client will need pay the additional costs.
- b. The contractor will need to pay the additional costs.
- c. The contractor will install it but not connect it to the supply.
- d. The client will have to install it themselves and take responsibility.

EXAM PRACTICE 5

EXAM PRACTICE 5: UNIT 005 UNDERSTAND HOW TO PLAN AND OVERSEE ELECTRICAL WORK ACTIVITIES

15 What is the **main** potential risk of having all materials delivered to a site before a six-month contract begins?

- a. Materials may be damaged through site activities.
- b. The materials needed may reduce in price.
- c. The regulations may change making materials unsuitable.
- d. The materials may become obsolete and additions are needed.

16 A site is located fifty miles from a contractor's office.

How can the risk of theft of plant be minimised cost effectively for this contract?

- a. Storing plant in a van overnight.
- b. Storing plant in the office each night.
- c. Storing the plant in a secure container within the site.
- d. Storing the plant under a groundsheet on the site.

Notes

Notes

Exam practice 5 questions with model answers: Unit 005 understand how to plan and oversee electrical work activities

1. What is the **most** suitable method of contacting a colleague to remind them, without too much disturbance late at night, that you will be late arriving on site the next day?

   - ○ a. Electronic mail.
   - ● b. Text message.
   - ○ c. Postal letter.
   - ○ d. Telephone call.

   Correct answer b
   A text message is a good method of contacting a person, regardless of the time, with the confirmation they have received it.

   Whilst email is a good way to communicate, urgent messages are not guaranteed due to spam filters. Post takes far too long and a telephone call late at night may not be well received by most people.

2. What is the most reliable method of communicating with other trades on a small site, to ensure all parties understand a minor problem that has occurred?

   - ● a. Face to face communication.
   - ○ b. Electronic mail (email).
   - ○ c. Social media messaging.
   - ○ d. Written posted letter.

   Correct answer a
   If it is a small site then talking to people is by far the best method of communication as you can ensure they have received the message and, through body language, understood it too.

3. How should you initially communicate with a work colleague who is lacking motivation due to personal issues?

   - ● a. Informal conversation.
   - ○ b. Sending a formal letter.
   - ○ c. Email with final written warning.
   - ○ d. Discussion via open social media.

   Correct answer a

EXAM PRACTICE 5

EXAM PRACTICE 5 QUESTIONS WITH MODEL ANSWERS: UNIT 005 UNDERSTAND HOW TO PLAN AND OVERSEE ELECTRICAL WORK ACTIVITIES

Notes

Having an informal conversation with a friend or colleague is the best way to help with personal issues. When you speak face to face you are able to pick up other messages such as body language and expressions.

4   What is the **most** appropriate method of ensuring many trades are up to date and understand major alterations to a contract?

- a. Text message.
- b. E-mail.
- ● c. Site meeting.
- d. Twitter.

Correct answer c
Holding regular site meetings is a good way to ensure every trade knows what is happening and are aware of alterations. Whilst the other methods may convey a message, implications cannot easily be discussed.

5   What card proves a person is fully qualified as an electrician?

- a. CSCS.
- ● b. ECS.
- c. PASMA.
- d. ID.

Correct answer b
The Electrotechnical Certification Scheme (ECS) card is the national scheme, run by the JIB, for individual electricians to prove competency. Only people who have completed all the recognised qualifications and demonstrated experience are eligible to gain the card, which shows they are fully qualified.

6   What is the **most** efficient method for selecting a shortlist of candidates, from many applications, for a job role?

- a. Hold interviews.
- b. Have trial periods.
- ● c. Request candidate CVs.
- d. Visit their current employers.

Correct answer c
When a shortlist from many candidates for a job is needed, the most efficient method of selecting the shortlist is to review individual Curriculum Vitae (CV) to see if applicants have the relevant skills required.

Notes

7   What needs to be considered when a work schedule is severely disrupted due to bad weather?

- ● a.   Availability of suitable labour.
- ○ b.   Availability of delivered materials.
- ○ c.   Suitability of specified accessories.
- ○ d.   Suitability of installed equipment.

Correct answer a
When a work schedule becomes disrupted, due to poor weather, labour requirements need to be reassessed as the labour may have been allocated to other sites.

8   What is the purpose of a variation order?

- ○ a.   To inform the client of amendments to the current Wiring Regulations.
- ● b.   To maintain a record of amendments to the original agreed contract.
- ○ c.   To allow persons on site to alter the contract to make work easier.
- ○ d.   To give a wholesaler permission to substitute items without discussion.

Correct answer b
When a change to the original agreed contract is made, a variation order is issued by the client to make those changes official and to keep a record of those agreed changes.

9   What document allows an organisation to check that delivered materials are in accordance to the original order?

- ○ a.   Variation order.
- ○ b.   Site diary.
- ● c.   Delivery note.
- ○ d.   Take-off sheets.

Correct answer c
The delivery note should be checked to the original order to ensure that items actually delivered are in accordance to those required by the original order.

EXAM PRACTICE 5

EXAM PRACTICE 5 QUESTIONS WITH MODEL ANSWERS: UNIT 005 UNDERSTAND HOW TO PLAN AND OVERSEE ELECTRICAL WORK ACTIVITIES

10 What document would be used to determine the quantity of accessories needed for a contract?

- ◉ a. Site plans.
- ○ b. Site diary.
- ○ c. Delivery note.
- ○ d. Critical network.

Correct answer a
Site plans would show what accessories are required for a site and would be used, in conjunction with a take-off sheet, to determine the quantities required.

11 What, from the following, is identified by a Gantt chart?

- ◉ a. Labour requirements at various stages of the contract.
- ○ b. Quantities of materials left over at the end of a contract.
- ○ c. How much material was taken to landfill or recycled.
- ○ d. The energy performance of an installed electrical system.

Correct answer a
Gantt charts indicate a programme of work and are used to determine labour and plant resources as they provide a visual guide to the stages of a contract.

12 What is the **main** purpose of a material schedule?

- ○ a. To ensure waste electrical materials are disposed of correctly.
- ○ b. To ensure materials are stored on site at the correct temperature.
- ◉ c. To ensure that the correct type of materials are purchased.
- ○ d. To ensure materials are handled in accordance with COSHH.

Correct answer c
A material schedule will provide information relating to the correct type of materials required by the client. This ensures they are fit for purpose and will indicate the quantity required. A client and contractor will agree a material schedule before work begins as this can affect the overall cost of the work and that the client's wishes are compliant with regulations.

Notes

Notes

13 What document should be agreed upon by the client and contractor, before work begins, to ensure all accessories are fit for purpose?

- ○ a. Variation order.
- ● b. Materials schedule.
- ○ c. Day-work sheets.
- ○ d. Delivery notes.

Correct answer b
A materials schedule will provide information relating to the correct type of materials required by the client. This ensures they are fit for purpose and will indicate the quantity required. A client and contractor will agree a material schedule before work begins as this can affect the overall cost of the work and that the client's wishes are compliant with regulations.

14 A client specifies that they want a particular socket-outlet outdoors which they do not realise is non-compliant. The correct type needed is much more expensive.

What is the **most** likely outcome if a contractor doesn't discuss this issue with the client before a contract is signed?

- ○ a. The client will need pay the additional costs.
- ● b. The contractor will need to pay the additional costs.
- ○ c. The contractor will install it but not connect it to the supply.
- ○ d. The client will have to install it themselves and take responsibility.

Correct answer b
Issues such as this must be discussed and agreed before a contract is signed. If the contractor doesn't identify potential non-compliances and agrees to the client's wishes, they are contractually liable for any costs required to make the work comply. This is why it is essential to survey and discuss all work before agreeing and commencing with a contract.

EXAM PRACTICE 5

EXAM PRACTICE 5 QUESTIONS WITH MODEL ANSWERS: UNIT 005 UNDERSTAND HOW TO PLAN AND OVERSEE ELECTRICAL WORK ACTIVITIES

Notes

15 What is the **main** potential risk of having all materials delivered to a site before a six-month contract begins?

- ⦿ a. Materials may be damaged through site activities.
- ○ b. The materials needed may reduce in price.
- ○ c. The regulations may change making materials unsuitable.
- ○ d. The materials may become obsolete and additions are needed.

Correct answer a
On a busy site, excessive materials which are needed much later on the site are likely to become damaged.

16 A site is located fifty miles from a contractor's office.

How can the risk of theft of plant be minimised cost effectively for this contract?

- ○ a. Storing plant in a van overnight.
- ○ b. Storing plant in the office each night.
- ⦿ c. Storing the plant in a secure container within the site.
- ○ d. Storing the plant under a groundsheet on the site.

Correct answer c
Although containers are expensive to hire, they are very secure. Many vans are broken into at night and taking plant that distance each night and collecting in the morning is not a cost effective use of labour.

Exam practice 5 answer key: Unit 005 understand how to plan and oversee electrical work activities

1 b	5 b	9 c	13 b
2 a	6 c	10 a	14 b
3 a	7 a	11 b	15 a
4 c	8 b	12 c	16 c

EXAM PRACTICE 6

Exam Practice 6

Exam practice 6: Unit 012 understand inspection, testing and commissioning   132

Exam practice 6 questions with model answers:
Unit 012 understand inspection, testing and commissioning   140

Exam practice 6 answer key: Unit 012 understand inspection, testing and commissioning   152

Exam practice 6: Unit 012 understand inspection, testing and commissioning

1. What does the Electricity at Work Regulations specifically require a person carrying out inspection and testing to be?

   - a. Competent.
   - b. Ordinary.
   - c. Responsible.
   - d. Instructed.

2. How must an isolated circuit be secured?

   - a. Using a warning notice.
   - b. With a lock having one key.
   - c. With a lock having many keys.
   - d. Using tape over the circuit breaker.

3. What does a proving unit do?

   - a. It checks that a circuit is safe to work on.
   - b. It checks that a circuit breaker is secured.
   - c. It simulates a supply to test the isolated circuit.
   - d. It simulates a supply to test the Voltage indicator.

4. Which test cannot be safely undertaken on a circuit that is **not** securely isolated?

   - a. Phase sequencing.
   - b. Insulation resistance.
   - c. Prospective fault current.
   - d. Earth fault loop impedance.

5. What must an inspector consider, for safety reasons, before isolating a supply?

   - a. Mobile phones are fully charged.
   - b. Lifts are not operating or occupied.
   - c. Emergency lights are fully discharged.
   - d. Fire alarms batteries are disconnected.

EXAM PRACTICE 6

EXAM PRACTICE 6: UNIT 012 UNDERSTAND INSPECTION, TESTING AND COMMISSIONING

6 What should never be undertaken unless a building is fully isolated?

- a. Remove the cover from a socket-outlet.
- b. Undertake earth fault loop impedance testing
- c. Disconnect main protective bonding conductors.
- d. Check the phase sequence of the supply.

7 Which IET Guidance Note is specific to the Initial Verification process?

- a. GN1
- b. GN3
- c. GN6
- d. GN8

8 Which of the following documents is statutory?

- a. EWR
- b. BS 7671
- c. HSE GS38
- d. GN2

9 What are the first three tests undertaken on a newly installed radial circuit?

- a. Continuity of CPC, Continuity of Ring, Insulation Resistance.
- b. Continuity of Ring, Insulation Resistance, Polarity.
- c. Continuity of Ring, Polarity, Earth Electrode Resistance.
- d. Continuity of CPC, Insulation Resistance, Polarity.

10 What test is undertaken at the origin of an installation?

- a. R_1
- b. Z_e
- c. r_n
- d. Z_s

Notes

11 What test is undertaken using a low resistance ohmmeter?

- a. Supply Polarity.
- b. Insulation Resistance.
- c. Circuit Polarity.
- d. Electrode resistance.

12 What is the unit of measurement displayed during an RCD functional test?

- a. milliseconds.
- b. milliamperes.
- c. milliohms.
- d. millivolts.

13 How would an insulation resistance tester be checked for function?

- a. Confirming the instrument on a known supply.
- b. Test with leads open then closed together.
- c. Test whilst holding the probe tips between fingers.
- d. Confirming the instrument's calibration date.

14 What document gives specific requirements for fuses in test leads that are used by electricians?

- a. BS 7671.
- b. IET GN4.
- c. ISO 9000.
- d. HSE GS38.

15 What procedure would need to be carried out, if test instruments were found to be very inaccurate, following accuracy checks?

- a. Installations tested since the last calibration would need re-testing.
- b. Installations tested since the last calibration must be shut-down.
- c. The HSE must be informed immediately using a RIDDOR form.
- d. The local authority must be informed so they can issue notices.

EXAM PRACTICE 6

EXAM PRACTICE 6: UNIT 012 UNDERSTAND INSPECTION, TESTING AND COMMISSIONING

16 What must be carried out if a test of insulation resistance produced a non-compliant result during the initial verification of a circuit?

- a. The fault must be repaired and live testing would be carried out.
- b. Testing would continue and the fault must be repaired after all tests.
- c. The fault must be repaired and start all testing from the beginning.
- d. The tests would continue and the problem is reported on the Certificate.

17 How are the continuity of Main Protective Bonding conductors verified when they are visible throughout their length?

- a. By visual inspection and checking tightness of connections.
- b. By testing their insulation resistance to a known earth.
- c. By testing earth fault loop impedance with them disconnected.
- d. By testing their voltage whilst introducing an earth fault.

18 What is verified when a result of R_1+R_2 is successfully recorded?

- a. That the circuit line and neutral is continuous in the circuit.
- b. That the line and CPC is continuous in the supply.
- c. That the neutral and CPC is continuous in the supply.
- d. That the line and CPC is continuous in the circuit.

19 Why is it necessary to test ring-final circuit continuity to establish that any unfused spurs are limited to one socket-outlet?

- a. As the cable used for the spur is not suitably sized for undervoltages.
- b. As the cable used for the spur is not suitably sized for overcurrent.
- c. As the cables wired in the ring will become un-balanced.
- d. As the cables wired in the ring will increase in resistance.

20 Why is it essential for a Main Protective Bonding Conductor, connected to a gas installation pipe, to have a low resistance?

- a. To ensure voltage appears on the gas pipe during an earth fault.
- b. To provide a path to earth if the gas pipe develops a leak.
- c. To ensure the circuit disconnects in the time specified by BS 7671.
- d. To ensure the circuit has a continuous line conductor.

Notes

Notes

21 What must be checked, with regards to light switches, to effectively test insulation resistance of a complete lighting circuit?

- a. Switches must be open.
- b. Switches must be disconnected.
- c. Switches must be closed.
- d. Switches must be isolated.

22 What is a possible reason for a low insulation resistance when testing a car park lighting circuit?

- a. Long cable lengths.
- b. Multiple circuits in parallel.
- c. Decreased cable CSA.
- d. A switch is in the off position.

23 What action, during the continuity of CPC (R_1+R_2) test for a lighting circuit, could prove correct polarity of that circuit?

- a. Energise the circuit.
- b. Operate light switches.
- c. Reverse test probes.
- d. Remove temporary link.

24 What test instrument can be used to confirm supply polarity is correct?

- a. Low resistance ohmmeter.
- b. Insulation resistance tester.
- c. Earth fault loop impedance tester.
- d. Earth electrode resistance tester.

25 What is the correct formula for calculating total earth fault loop impedance?

- a. $Z_e = Z_s + R_1 + R_2$
- b. $Z_s = Z_e + R_1 + R_2$
- c. $Z_e = Z_s \times R_1 + R_2$
- d. $Z_s = Z_{es} \times R_1 + R_2$

EXAM PRACTICE 6

EXAM PRACTICE 6: UNIT 012 UNDERSTAND INSPECTION, TESTING AND COMMISSIONING

26 What would be recorded as the Prospective Fault Current (I_{pf}) for a single-phase installation where the following test results were obtained?

- PEFC = 0.5 kA
- PSCC = 0.9 kA

- a. 0.5 kA
- b. 0.9 kA
- c. 1.4 kA
- d. 1.8 kA

27 What is an alternative method to direct measurement, for obtaining the Prospective Fault Current for an installation?

- a. Multiplying the maximum Z_s by 2.
- b. Using the maximum demand value.
- c. Recording the breaking capacity of the Distribution Network Operator (DNO) fuse.
- d. Enquiry to the Distribution Network Operator.

28 What characteristic of a circuit breaker, close to the origin, is checked against the I_{pf} for an installation?

- a. Nominal rating.
- b. Disconnection time.
- c. Short circuit capacity.
- d. Overload current.

29 What is the maximum disconnection time for a 30 mA RCD when tested using 150 mA?

- a. 30 ms.
- b. 40 ms.
- c. 200 ms.
- d. 300 ms.

Notes

30 What is the main risk to three-phase machines if phase-sequencing is incorrect?

- a. Machines may run backwards.
- b. Machines will run too slow.
- c. Machines will overload.
- d. Machines will run too fast.

31 What instrument is used to verify phase-sequence?

- a. A three-lead single-phase approved voltmeter.
- b. A two-lead continuity test set.
- c. A three-lead rotating disc instrument.
- d. A two-lead proving unit.

32 What functional test is carried out on an RCD?

- a. Confirmation that the RCD is 30 mA.
- b. Confirmation of the test button operation.
- c. Confirmation that the RCD is Type A.
- d. Confirmation of the fly lead connection.

33 What must a client be made aware of following the verification of an installation having a range of RCBOs?

- a. The need to power off the installation to allow the RCBOs to cool once a year.
- b. The need to have RCBOs replaced if they trip more than once.
- c. The need to operate the test button on a regular basis.
- d. The need to test the RCBOs yearly with a test instrument.

34 Which document from BS 7671 can only be used to verify an addition or alteration to an existing circuit only?

- a. Electrical installation Condition Report.
- b. Minor Electrical Installation Works Certificate.
- c. Electrical Installation Certificate.
- d. Schedule of Test Results.

35 What information, relating to the supply conductors, would be ticked for a single-phase installation?

- a. 1-phase: 2-wire.
- b. 1-phase: 3-wire.
- c. 3-phase: 2-wire.
- d. 3-phase: 3-wire.

Notes

Exam practice 6 questions with model answers: Unit 012 understand inspection, testing and commissioning

1. What does the Electricity at Work Regulations specifically require a person carrying out inspection and testing to be?

   - ◉ a. Competent.
   - ○ b. Ordinary.
   - ○ c. Responsible.
   - ○ d. Instructed.

   Correct answer a
   Any person undertaking electrical work must be competent. This means they must have sufficient knowledge and experience of the work undertaken.

2. How must an isolated circuit be secured?

   - ○ a. Using a warning notice.
   - ◉ b. With a lock having one key.
   - ○ c. With a lock having many keys.
   - ○ d. Using tape over the circuit breaker.

   Correct answer b
   An isolated circuit must be locked in the off position using a lock having one key. This key would be kept by the person working on the circuit at all times.

3. What does a proving unit do?

   - ○ a. It checks that a circuit is safe to work on.
   - ○ b. It checks that a circuit breaker is secured.
   - ○ c. It simulates a supply to test the isolated circuit.
   - ◉ d. It simulates a supply to test the Voltage indicator.

   Correct answer d
   A proving unit is used to check that approved voltage indicators function correctly. Indicators must be proven to function properly either before or after checking that a circuit is dead.

EXAM PRACTICE 6

EXAM PRACTICE 6 QUESTIONS WITH MODEL ANSWERS: UNIT 012 UNDERSTAND INSPECTION, TESTING AND COMMISSIONING

4 Which test cannot be safely undertaken on a circuit that is **not** securely isolated?

- a. Phase sequencing.
- ● b. Insulation resistance.
- c. Prospective fault current.
- d. Earth fault loop impedance.

Correct answer b
Insulation resistance testing cannot be safely undertaken unless the circuit is securely isolated from the supply.

5 What must an inspector consider, for safety reasons, before isolating a supply?

- a. Mobile phones are fully charged.
- ● b. Lifts are not operating or occupied.
- c. Emergency lights are fully discharged.
- d. Fire alarms batteries are disconnected.

Correct answer b
For the safety of others, lifts in a building must be empty before a supply is isolated as persons may become trapped once the supply is off.

6 What should never be undertaken unless a building is fully isolated?

- a. Remove the cover from a socket-outlet.
- b. Undertake earth fault loop impedance testing
- ● c. Disconnect main protective bonding conductors.
- d. Check the phase sequence of the supply.

Correct answer c
The main protective bonding within a building, such as those connecting gas and water utility pipework, must never be disconnected unless the supply is fully isolated.

Notes

Notes

7 Which IET Guidance Note is specific to the Initial Verification process?

- a. GN1
- b. GN3 *(selected)*
- c. GN6
- d. GN8

Correct answer b
IET Guidance Note 3 specifically covers all aspects of inspection and testing including the Initial Verification process.

8 Which of the following documents is statutory?

- a. EWR *(selected)*
- b. BS 7671
- c. HSE GS38
- d. GN2

Correct answer a
EWR or The Electricity at Work Regulations is a statutory document which means there is a legal duty to comply with it. Non-statutory documents such as HSE GS38 will often give guidance on how to comply with statutory documents.

9 What are the first three tests undertaken on a newly installed radial circuit?

- a. Continuity of CPC, Continuity of Ring, Insulation Resistance.
- b. Continuity of Ring, Insulation Resistance, Polarity.
- c. Continuity of Ring, Polarity, Earth Electrode Resistance.
- d. Continuity of CPC, Insulation Resistance, Polarity. *(selected)*

Correct answer d
The test, Continuity of ring, does not apply to radial circuits so the first three tests that should be undertaken, in accordance with BS 7671 are

- Continuity of CPC.
- Insulation Resistance.
- Polarity.

BS 7671 also states that those tests must be undertaken in the order given.

EXAM PRACTICE 6

EXAM PRACTICE 6 QUESTIONS WITH MODEL ANSWERS: UNIT 012 UNDERSTAND INSPECTION, TESTING AND COMMISSIONING

10 What test is undertaken at the origin of an installation?

- a. R_1
- ● b. Z_e
- c. r_n
- d. Z_s

Correct answer b
The earth fault loop impedance external to the installation (Z_e) is carried out at the origin of the installation. All the other values are in relation to circuits within the installation.

11 What test is undertaken using a low resistance ohmmeter?

- a. Supply Polarity.
- b. Insulation Resistance.
- ● c. Circuit Polarity.
- d. Electrode resistance.

Correct answer c
The polarity test to circuits within an installation is carried out using a low resistance ohmmeter, usually during continuity testing.

12 What is the unit of measurement displayed during an RCD functional test?

- ● a. milliseconds.
- b. milliamperes.
- c. milliohms.
- d. millivolts.

Correct answer a
An RCD test creates a fault at a particular current and measures how quickly the RCD trips in milliseconds.

Notes

Notes

13 How would an insulation resistance tester be checked for function?

- a. Confirming the instrument on a known supply.
- **b. Test with leads open then closed together.** ●
- c. Test whilst holding the probe tips between fingers.
- d. Confirming the instrument's calibration date.

Correct answer b
The instrument should be tested for function before use by testing with the probes apart (open circuit) then with the probes touching (closed circuit) to check the meter gives a high then low reading respectively.

14 What document gives specific requirements for fuses in test leads that are used by electricians?

- a. BS 7671.
- b. IET GN4.
- c. ISO 9000.
- **d. HSE GS38.** ●

Correct answer d
HSE GS38 is titled 'Electrical test equipment for use on low voltage electrical systems' and gives specific requirements for fuses in test leads amongst other requirements. It is essential, for reasons of safety, that electricians using test equipment are familiar with GS 38 before any testing.

15 What procedure would need to be carried out, if test instruments were found to be very inaccurate, following accuracy checks?

- **a. Installations tested since the last calibration would need re-testing.** ●
- b. Installations tested since the last calibration must be shut-down.
- c. The HSE must be informed immediately using a RIDDOR form.
- d. The local authority must be informed so they can issue notices.

Correct answer a
If an instrument is found to be inaccurate, any installation tested by that instrument should be retested to see if the instrument was accurate during that inspection and test and that the results obtained were reliable. For this reason, it is essential that instruments are checked for accuracy at frequent intervals.

EXAM PRACTICE 6

EXAM PRACTICE 6 QUESTIONS WITH MODEL ANSWERS: UNIT 012 UNDERSTAND INSPECTION, TESTING AND COMMISSIONING

16 What must be carried out if a test of insulation resistance produced a non-compliant result during the initial verification of a circuit?

- a. The fault must be repaired and live testing would be carried out.
- b. Testing would continue and the fault must be repaired after all tests.
- ● c. The fault must be repaired and start all testing from the beginning.
- d. The tests would continue and the problem is reported on the Certificate

Correct answer c
The fault on the circuit must be repaired and testing of that circuit must start again as the fault may have influenced some of the previous results.

17 How are the continuity of Main Protective Bonding conductors verified when they are visible throughout their length?

- ● a. By visual inspection and checking tightness of connections.
- b. By testing their insulation resistance to a known earth.
- c. By testing earth fault loop impedance with them disconnected.
- d. By testing their voltage whilst introducing an earth fault.

Correct answer a
If the main protective Bonding Conductors are visible throughout their length, they may be verified by visual inspection and the effectiveness of their connections verified by touch.

18 What is verified when a result of R_1+R_2 is successfully recorded?

- a. That the circuit line and neutral is continuous in the circuit.
- b. That the line and CPC is continuous in the supply.
- c. That the neutral and CPC is continuous in the supply.
- ● d. That the line and CPC is continuous in the circuit.

Correct answer d
When R_1+R_2 is verified, the test is carried out between a circuit line and CPC to ensure they are continuous throughout the circuit.

Notes

19 Why is it necessary to test ring-final circuit continuity to establish that any unfused spurs are limited to one socket-outlet?

- a. As the cable used for the spur is not suitably sized for undervoltages.
- ● b. As the cable used for the spur is not suitably sized for overcurrent.
- c. As the cables wired in the ring will become un-balanced.
- d. As the cables wired in the ring will increase in resistance.

Correct answer b
If too many socket-outlets are connected to any un-fused spur off the ring-final circuit, the cable will not be suitably protected by the circuit overcurrent device.

20 Why is it essential for a Main Protective Bonding Conductor, connected to a gas installation pipe, to have a low resistance?

- ● a. To ensure voltage appears on the gas pipe during an earth fault.
- b. To provide a path to earth if the gas pipe develops a leak.
- c. To ensure the circuit disconnects in the time specified by BS 7671.
- d. To ensure the circuit has a continuous line conductor.

Correct answer a
The purpose of Main Protective Bonding is to ensure that any voltages present on earthed parts also appears on extraneous parts such as the gas pipe. This will create equal potential in the event of a fault.

21 What must be checked, with regards to light switches, to effectively test insulation resistance of a complete lighting circuit?

- a. Switches must be open.
- b. Switches must be disconnected.
- ● c. Switches must be closed.
- d. Switches must be isolated.

Correct answer c
All switches must be closed to obtain a reading through the entire circuit.

EXAM PRACTICE 6

EXAM PRACTICE 6 QUESTIONS WITH MODEL ANSWERS: UNIT 012 UNDERSTAND INSPECTION, TESTING AND COMMISSIONING

Notes

22 What is a possible reason for a low insulation resistance when testing a car park lighting circuit?

- ● a. Long cable lengths.
- ○ b. Multiple circuits in parallel.
- ○ c. Decreased cable CSA.
- ○ d. A switch is in the off position.

Correct answer a
Where a circuit has a long cable run, the insulation resistance could reduce to a lower than expected value. The longer the cable, the more insulation is in parallel reducing the overall value.

23 What action, during the continuity of CPC (R_1+R_2) test for a lighting circuit, could prove correct polarity of that circuit?

- ○ a. Energise the circuit.
- ● b. Operate light switches.
- ○ c. Reverse test probes.
- ○ d. Remove temporary link.

Correct answer b
By operating the light switches during the test and seeing the reading open circuit is a method of verifying polarity for that circuit.

24 What test instrument can be used to confirm supply polarity is correct?

- ○ a. Low resistance ohmmeter.
- ○ b. Insulation resistance tester.
- ● c. Earth fault loop impedance tester.
- ○ d. Earth electrode resistance tester.

Correct answer c
An earth fault loop impedance tester also has a polarity indicator which verifies the supply polarity when the instrument is correctly connected to a live supply.

Notes

25 What is the correct formula for calculating total earth fault loop impedance?

- a. $Z_e = Z_s + R_1 + R_2$
- ● b. $Z_s = Z_e + R_1 + R_2$
- c. $Z_e = Z_s \times R_1 + R_2$
- d. $Z_s = Z_s \times R_1 + R_2$

Correct answer b
The total earth fault loop impedance (Z_s) is calculated by adding together the external loop impedance (Z_e) to the circuit line and CPC resistance (R_1+R_2).

26 What would be recorded as the Prospective Fault Current (I_{pf}) for a single-phase installation where the following test results were obtained?

- PEFC = 0.5 kA
- PSCC = 0.9 kA

- a. 0.5 kA
- ● b. 0.9 kA
- c. 1.4 kA
- d. 1.8 kA

Correct answer b
For a single-phase installation, the largest value from the tests of PEFC and PSCC is recorded as the Ipf for the installation.

27 What is an alternative method to direct measurement, for obtaining the Prospective Fault Current for an installation?

- a. Multiplying the maximum Z_s by 2.
- b. Using the maximum demand value.
- c. Recording the breaking capacity of the Distribution Network Operator (DNO) fuse.
- ● d. Enquiry to the Distribution Network Operator.

Correct answer d
If the I_{pf} cannot be measured, a suitable alternative would be to make an enquiry to the DNO.

EXAM PRACTICE 6

EXAM PRACTICE 6 QUESTIONS WITH MODEL ANSWERS: UNIT 012 UNDERSTAND INSPECTION, TESTING AND COMMISSIONING

28 What characteristic of a circuit breaker, close to the origin, is checked against the I_{pf} for an installation?

- ○ a. Nominal rating.
- ○ b. Disconnection time.
- ● c. Short circuit capacity.
- ○ d. Overload current.

Correct answer c
The purpose of the Prospective Fault Current test is to ensure the short circuit capacity of circuit breakers, installed at the origin of an installation, are suitable for any faults.

29 What is the maximum disconnection time for a 30 mA RCD when tested using 150 mA?

- ○ a. 30 ms.
- ● b. 40 ms.
- ○ c. 200 ms.
- ○ d. 300 ms.

Correct answer b
Where a 30 mA RCD is installed, the disconnection time when tested at x5 $I_{\Delta N}$ should not exceed 40 ms in accordance with performance values given in Appendix 3 of BS 7671.

30 What is the main risk to three-phase machines if phase-sequencing is incorrect?

- ● a. Machines may run backwards.
- ○ b. Machines will run too slow.
- ○ c. Machines will overload.
- ○ d. Machines will run too fast.

Correct answer a
If the phase sequence is incorrect, three-phase machines will rotate in the opposite direction.

Notes

31 What instrument is used to verify phase-sequence?

- a. A three-lead single-phase approved voltmeter.
- b. A two-lead continuity test set.
- **c. A three-lead rotating disc instrument.**
- d. A two-lead proving unit.

Correct answer c
One type of phase sequence instrument is a small three phase motor which causes a disc to rotate in a direction specific to the phase sequence.

32 What functional test is carried out on an RCD?

- a. Confirmation that the RCD is 30 mA.
- **b. Confirmation of the test button operation.**
- c. Confirmation that the RCD is Type A.
- d. Confirmation of the fly lead connection.

Correct answer b
One functional test which is carried out on an RCD is that the test button, when operated, causes disconnection of the RCD.

33 What must a client be made aware of following the verification of an installation having a range of RCBOs?

- a. The need to power off the installation to allow the RCBOs to cool once a year.
- b. The need to have RCBOs replaced if they trip more than once.
- **c. The need to operate the test button on a regular basis.**
- d. The need to test the RCBOs yearly with a test instrument.

Correct answer c
In accordance with the test notice required on a distribution board containing RCDs/RCBOs, the devices should be tested quarterly by operating the test button. This ensures the mechanism remains free.

EXAM PRACTICE 6

EXAM PRACTICE 6 QUESTIONS WITH MODEL ANSWERS: UNIT 012 UNDERSTAND INSPECTION, TESTING AND COMMISSIONING

34 Which document from BS 7671 can only be used to verify an addition or alteration to an existing circuit only?

- a. Electrical installation Condition Report.
- b. Minor Electrical Installation Works Certificate.
- c. Electrical Installation Certificate.
- d. Schedule of Test Results.

Correct answer b

The document used to verify the addition or alteration to an existing circuit is the Minor Electrical Installation Works Certificate, which is its sole purpose. Whilst an Electrical Installation Certificate can also be used for this purpose, it is also used for other situations such as a complete new installation.

35 What information, relating to the supply conductors, would be ticked for a single-phase installation?

- a. 1-phase: 2-wire.
- b. 1-phase: 4-wire.
- c. 3-phase: 2-wire.
- d. 3-phase: 4-wire.

Correct answer a

If the electrical installation is single-phase, as detailed in the question, it can only have a 2-wire supply conductor arrangement which is line and neutral. The earthing conductor is not considered in this situation.

Notes

Exam practice 6 answer key: Unit 012 understand inspection, testing and commissioning

1 a	8 a	15 a	22 a	29 b
2 b	9 d	16 c	23 b	30 a
3 d	10 b	17 a	24 c	31 c
4 b	11 c	18 d	25 b	32 b
5 b	12 a	19 b	26 b	33 c
6 c	13 b	20 a	27 d	34 b
7 b	14 d	21 c	28 c	35 a

Exam Practice 7

Exam Practice 7

Exam practice 7: Unit 112 written paper understand inspection, testing and commissioning   154

Exam practice 7 questions with model answers: Unit 112 written paper understand inspection, testing and commissioning   158

Exam practice 7: Unit 112 written paper understand inspection, testing and commissioning

Section A of the exam:

1. a) State **three** responsibilities of the inspector when carrying out an inspection and test of an electrical installation. (3 marks)
   b) State the titles of the **three** documents that must be handed to the client on completion of an initial verification. (3 marks)
   c) State, on which document in 1b) above, the verification of continuity of main protective bonding conductors is recorded. (1 mark)
   d) List **four** items of information that must be recorded in the "Supply Characteristics and Earthing Arrangements" section of the initial verification documentation. (4 marks)
   e) The documentation for initial verification includes a section entitled "Comments on Existing Installation".
      i) State, under what circumstances, this section would contain a comment referring to the condition of the existing installation. (2 marks)
      ii) State an example of a comment that may be entered in this section of the document. (2 marks)

EXAM PRACTICE 7

EXAM PRACTICE 7: UNIT 112 WRITTEN PAPER UNDERSTAND INSPECTION, TESTING AND COMMISSIONING

2   A newly installed socket-outlet circuit in a domestic premise is to be inspected.

State, using the grid below, **five** items which will be inspected, **excluding** the wiring and terminations, **one** check to be carried out on **each** item and the human sense to be used in **each** case. (15 marks)

Item no	Item to be inspected	Checked to confirm	Human sense used
1			
2			
3			
4			
5			

3   a) The HSE publish a guidance that relates to the requirements for electrical test equipment.
   i) State the title of this guidance. (1 mark)
   ii) State whether this document is statutory or non-statutory. (1 mark)
   iii) State **three** requirements for test leads given in this document. (3 marks)
   b) Describe, in detail, the procedure to be followed when carrying out safe isolation of a single-phase distribution board on an existing installation. Permission has already been given. (10 marks)

Notes

Section B of the exam:

All questions carry equal marks. Answer ALL THREE questions. Show ALL calculations.

Questions 4 to 6 all refer to the scenario below. Ensure you read this scenario before attempting these questions. Answers you provide must reflect the detail and information given in the scenario.

Scenario

A new small workshop unit with offices has been constructed. The electrical installation is supplied from a 400/230 V TN-C-S system. A metal-clad three-phase distribution board supplies power and lighting to the workshop and the offices.

A three-phase socket-outlet with a local switch is located in the workshop. The circuit is wired in four-core 70 °C thermoplastic insulated steel-wired armoured cable, protected by a three-phase 16 A BS EN 60898 circuit breaker.

The single-phase general power and lighting circuits for the whole unit are wired in 70 °C thermoplastic single-core cables, with copper conductors, installed in surface mounted steel trunking and conduit. The 13 A socket-outlet circuits are protected by BS EN 61009 RCBOs. The lighting circuits are protected by BS EN 60898 circuit breakers.

Outside lighting includes a PIR motion detector. Inside lighting includes two-way lighting in the offices and workshop.

Testing is to be carried out at an ambient temperature of 20 °C.

EXAM PRACTICE 7

EXAM PRACTICE 7: UNIT 112 WRITTEN PAPER UNDERSTAND INSPECTION, TESTING AND COMMISSIONING

4 An insulation resistance test is to be carried out on the whole installation.

   a) i) State the instrument to be used. (1 mark)
      ii) State the test voltage to be applied during the test. (1 mark)
      iii) State the minimum acceptable insulation resistance test value. (1 mark)
      iv) State **one** check to be taken regarding the test leads before the test is carried out. (1 mark)
   b) Describe, in detail, how the insulation resistance test is to be carried out on the whole installation. The information given in 4a) above need not be repeated here. (8 marks)
   c) State the three actions to be taken if an unsatisfactory test result is obtained when correctly carrying out the insulation resistance test in 4b) above. (3 marks)

5 a) An earth fault loop impedance test is to be carried out on the final circuit supplying the three-phase socket-outlet in the workshop.
      i) State the test instrument to be used for this test. (1 mark)
      ii) State **three** checks to be carried out on the test instrument before conducting the test. (3 marks)
      iii) State whether the Main Protective Bonding Conductors are connected or disconnected from the MET during the test. (1 marks)
   b) Describe, in detail, the procedure for carrying out the earth fault loop impedance test on the three-phase socket-outlet circuit. The information given in 5 a) above need not be repeated here. (10 marks)

6 Describe, with the aid of a fully labelled diagram, the earth fault loop path for one of the lighting circuits. (15 marks)

Notes

Notes

Exam practice 7 questions with model answers: Unit 112 written paper understand inspection, testing and commissioning

1. a) State three responsibilities of the inspector when carrying out an inspection and test of an electrical installation. (3 marks)
   b) State the titles of the three documents that must be handed to the client on completion of an initial verification. (3 marks)
   c) State, on which document in 1b) above, the verification of continuity of main protective bonding conductors is recorded. (1 mark)
   d) List four items of information that must be recorded in the "Supply Characteristics and Earthing Arrangements" section of the initial verification documentation. (4 marks)
   e) The documentation for initial verification includes a section entitled "Comments on Existing Installation".
      i) State, under what circumstances, this section would contain a comment referring to the condition of the existing installation. (2 marks)
      ii) State an example of a comment that may be entered in this section of the document. (2 marks)

Model answer:

1a) Any three from the following or similar wording
- No danger to persons/livestock
- Property not damaged
- Compare results with design criteria
- Compare results with BS 7671 requirements
- Confirm compliance
- Take a view on condition

Tip: Only the first three answers will be accepted.

1b) Correct titles must be used for the following three documents
- Electrical Installation Certificate
- Schedule of Inspections
- Schedule of Test Results

EXAM PRACTICE 7

EXAM PRACTICE 7 QUESTIONS WITH MODEL ANSWERS: UNIT 112 WRITTEN PAPER: UNDERSTAND INSPECTION, TESTING AND COMMISSIONING

1c) The Electrical Installation Certificate (this does not need to be a correct title if referenced to, e.g. 'the certificate').

1d) Any four from the following are acceptable:
- Earthing arrangement (system)
- Number and type of live conductors/AC/DC/single phase/three phase
- Voltage
- Frequency
- PFC
- Z_e
- Supply protective device – BS/type
- Supply protective device – rating or size

1e) similar wording may be used for

i) During inspection and test of an **addition/alteration** (1) to **an existing installation** (1).

ii) **Existing** (1) installation in good **condition** (1).

2 A newly installed socket-outlet circuit in a domestic premise is to be inspected.

State, using the grid below, five items which will be inspected, excluding the wiring and terminations, one check to be carried out on each item and the human sense to be used in each case. (15 marks)

Model answer:

Any five relevant items such as the following examples could be placed in the table as shown below:

1 mark per item
1 mark per check
1 mark per sense

Item no	Item to be inspected	Checked to confirm	Human sense used
1	Socket-outlets	to appropriate BS	Sight
2	Socket-outlets	correct position/height	Sight
3	Socket-outlets	suitable for environment	Sight
4	Socket-outlets	not damaged	Sight
5	Socket-outlets	securely fixed	Sight
6	Circuit protection	correct rating/size	Sight
7	Additional protection	30 mA RCD	Sight
8	Cable entry	protected	Sight

3 a) The HSE publish a guidance that relates to the requirements for electrical test equipment.
  i) State the title of this guidance. (1 mark)
  ii) State whether this document is statutory or non-statutory. (1 mark)
  iii) State **three** requirements for test leads given in this document. (3 marks)
 b) Describe, in detail, the procedure to be followed when carrying out safe isolation of a single-phase distribution board on an existing installation. Permission has already been given. (10 marks)

Model answer:

3a)

  i) GS38 (Full title may be given)
  ii) Non-statutory
  iii) Any three from:
   - Suitably Insulated
   - Sheathed against damage
   - Colour coded
   - Flexible
   - Suitable length (not too long, not too short)
   - No accessible exposed conductors (ie shrouded connectors)
   - Minimal exposed tips (2-4 mm)

EXAM PRACTICE 7

EXAM PRACTICE 7 QUESTIONS WITH MODEL ANSWERS: UNIT 112 WRITTEN PAPER: UNDERSTAND INSPECTION, TESTING AND COMMISSIONING

3b) Procedure to contain the following stages. Order is not critical for blue items

- Isolate main switch **or** isolate supply
- Lock off
- Fit sign
- Retain key
- Select voltage indicator **or** voltage tester
- GS 38 compliant
- Test voltage indicator on known supply or unit
- Test Live to Live (or L-N)
- Test Lives to earth (or LN-E)
- Or test between Live conductors and Live conductors to earth
- Test voltage indicator after use

1 mark per stage

Notes

Section B of the exam: All questions carry equal marks. Answer ALL THREE questions. Show ALL calculations.

Questions 4 to 6 all refer to the scenario. Ensure you read this scenario before attempting these questions. Answers you provide must reflect the detail and information given in the scenario.

Scenario

A new small workshop unit with offices has been constructed. The electrical installation is supplied from a 400/230 V TN-C-S system. A metal-clad three-phase distribution board supplies power and lighting to the workshop and the offices.

A three-phase socket-outlet with a local switch is located in the workshop. The circuit is wired in four-core 70 °C thermoplastic insulated steel-wired armoured cable, protected by a three-phase 16 A BS EN 60898 circuit breaker.

The single-phase general power and lighting circuits for the whole unit are wired in 70 °C thermoplastic single-core cables, with copper conductors, installed in surface mounted steel trunking and conduit. The 13 A socket-outlet circuits are protected by BS EN 61009 RCBOs. The lighting circuits are protected by BS EN 60898 circuit breakers.

Outside lighting includes a PIR motion detector. Inside lighting includes two-way lighting in the offices and workshop.

Testing is to be carried out at an ambient temperature of 20 °C.

Notes

4 An insulation resistance test is to be carried out on the whole installation.

a) i)   State the instrument to be used. (1 mark)
    ii) State the test voltage to be applied during the test. (1 mark)
    iii) State the minimum acceptable insulation resistance test value. (1 mark)
    iv) State **one** check to be taken regarding the test leads before the test is carried out. (1 mark)

b) Describe, in detail, how the insulation resistance test is to be carried out on the whole installation. The information given in 4a) above need not be repeated here. (8 marks)

c) State the **three** actions to be taken if an unsatisfactory test result is obtained when correctly carrying out the insulation resistance test in 4b) above. (3 marks)

Model answer:

4a)

   i)   Insulation resistance tester
   ii) 500 V
   iii) 1 MΩ
   iv) Must be to GS 38 (1) **OR** continuous/unbroken (1)

4b) 1 mark for each of the following steps (similar wording):

- Remove loads (lamps, plugged in appliances, pilot lights, PIR etc.) (1)
- Link in PIR the switched line conductor (1)
- Close CBs, switches, isolators etc. (1)
- Test between live conductors (1)
- Test between live conductors and Earth (1)
- Repeat for 2-way switching (1)
- Record results (1)
- Reinstate loads and switches (1)

4c) Answer to include the following or similar wording

- Locate the problem
- Fix the problem
- Re-test necessary tests

EXAM PRACTICE 7

EXAM PRACTICE 7 QUESTIONS WITH MODEL ANSWERS: UNIT 112 WRITTEN PAPER: UNDERSTAND INSPECTION, TESTING AND COMMISSIONING

5 a) An earth fault loop impedance test is to be carried out on the final circuit supplying the three-phase socket-outlet in the workshop.

    i) State the test instrument to be used for this test. (1 mark)
    ii) State **three** checks to be carried out on the test instrument before conducting the test. (3 marks)
    iii) State whether the Main Protective Bonding Conductors are connected or disconnected from the MET during the test. (1 mark)

b) Describe, in detail, the procedure for carrying out the earth fault loop impedance test on the three-phase socket-outlet circuit. The information given in 5 a) above need not be repeated here. (10 marks)

Model answer:

5a)
    i) Earth fault loop impedance tester
    ii) Within calibration
       Not damaged
       Functions correctly
       Battery condition sufficient
    iii) Connected

5b) The procedure is:
- Check power is off (1)
- Access socket-outlet terminals (1) **OR** use plug adaptor (3)
- Energise circuit (1)
- Use GS 38 instrument/leads/probes (1)
- Measure Z_s between L1-E/cpc (1)
- Repeat for L2-E/cpc and L3-E/cpc (1)
- Record (1) highest result (1) **OR** record all three results (2)
- Power off (1)
- Reinstate (1) **OR** remove adaptor (2)

Notes

6   Describe, with the aid of a fully labelled diagram, the earth fault loop path for one of the lighting circuits. (15 marks)

Model answer:

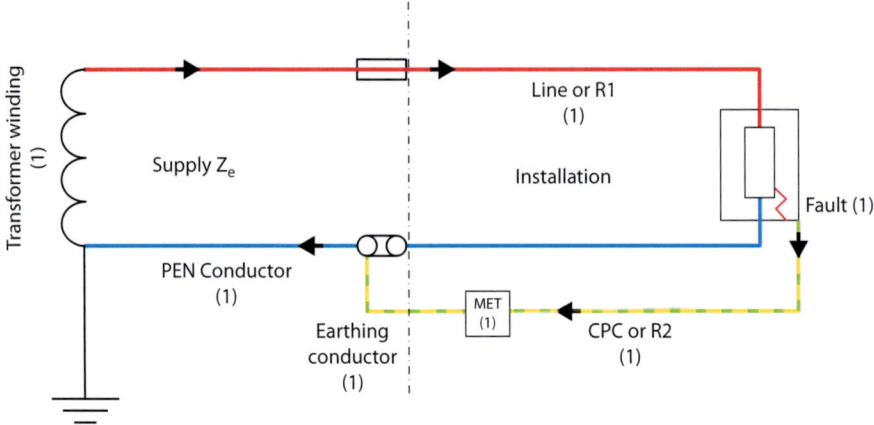

Marking is typically allocated for

- System diagram = 4 marks
- Path clearly shown = 4 marks
- Labels = 7 marks

If the wrong system or incomplete system so the type of system cannot be identified is drawn – 0 marks

If the fault path is not shown returning to, and passing through, the transformer winding maximum 2 marks for the path. (Candidate must show fault path through the star point and into the transformer winding by whatever means.)

If the path is incomplete – 2 marks max

Labels marks only for labelling shown – 1 mark each

Exam Practice 8

Exam Practice 8

Exam practice 8: Unit 014 understand fault diagnosis and rectification 166

Exam practice 8 questions with answers: Unit 014 understand fault diagnosis and rectification 173

Exam practice 8 answer key: Unit 014 understand fault diagnosis and rectification 184

Exam practice 8: Unit 014 understand fault diagnosis and rectification

1. What is the **most** likely reason an electrician received an electric shock whilst undertaking fault diagnosis?

   - a. No 30 mA RCD protection was present.
   - b. Safe isolation was not undertaken.
   - c. The fault tripped the circuit breaker.
   - d. The correct PPE was not present.

2. What is the final method of protection when considering a risk assessment?

   - a. Safe and secure isolation.
   - b. Personal protective equipment.
   - c. Warning notices and signage.
   - d. Communication with personnel.

3. What are the minimum voltages where GS38 becomes applicable?

   - a. 50 V DC: 50 V AC
   - b. 50 V DC: 120 V AC
   - c. 120 V DC: 50 V AC
   - d. 120 V DC: 120 V AC

4. What is the **most** useful document when establishing the correct operation of a specific item of equipment?

   - a. As fitted drawings and plans.
   - b. Electrical Installation Certificate.
   - c. Minor Electrical Works Certificate.
   - d. Manufacturers' instruction manual.

EXAM PRACTICE 8

EXAM PRACTICE 8: UNIT 014 UNDERSTAND FAULT DIAGNOSIS AND RECTIFICATION

5. What is the **main** advantage for communicating with users before undertaking fault diagnosis work to equipment?

   - a. To ascertain the nature of the fault.
   - b. To establish the final cost of the repair work.
   - c. To ascertain the location of catering facilities.
   - d. To establish how to repair the equipment.

6. What is the **most** likely reason for the continuous operation of an RCBO protecting a lighting circuit within a building being refurbished?

   - a. Overloading of the circuit.
   - b. Damage to a concealed cable.
   - c. Corrosion in a light switch.
   - d. Loose earth connection in light.

7. What is the **most** likely reason for an arcing sound in a light switch?

   - a. Voltage drop.
   - b. Open circuit.
   - c. Short circuit.
   - d. Loose connection.

8. What causes a reduced voltage at a load?

   - a. Short circuit lengths.
   - b. Open circuit faults.
   - c. Long circuit lengths.
   - d. Short circuit faults.

9. What is the **most** likely cause when a circuit breaker operates, is re-set, then operates again after several minutes?

   - a. Voltage drop.
   - b. Overloading.
   - c. Short circuit.
   - d. Open circuit.

Notes

10

Line-Line loop	0.4 Ω	Step 1 results
Neutral-Neutral loop	0.4 Ω	
CPC-CPC loop	0.4 Ω	
Readings at socket-outlet front	Open circuit	Step 2 results **Line-Neutral**
	0.2 Ω	Step 3 results **Line-CPC**

A continuity of ring-final circuit test is carried out.

What is the **most** likely fault at a socket-outlet given the readings in the table above?

- a. Reversed polarity between Line and CPC.
- b. A break in the line and the Neutral.
- c. Reversed polarity between line and neutral.
- d. A break in the neutral conductor.

11 What is the **most** likely reason for an RCD, controlling socket-outlets, to trip intermittently in a dwelling?

- a. A short in the wiring system.
- b. A faulty appliance such as a fridge.
- c. An open circuit in the wiring system.
- d. A loose bonding connection at the installation MET.

12 What location would create the greatest risk of electrocution due to a reduction in body resistance?

- a. Working on a wooden platform.
- b. Working in a metal ducting system.
- c. Working in a tiled bathroom.
- d. Working on an asphalt roof.

EXAM PRACTICE 8

EXAM PRACTICE 8: UNIT 014 UNDERSTAND FAULT DIAGNOSIS AND RECTIFICATION

Notes

13 What equipment is most vulnerable to damage by insulation resistance testing?

- a. Discharge lamps.
- b. Circuit wiring.
- c. Electronic controllers.
- d. Circuit breakers.

14 What is the **main** risk when working on isolated equipment having power factor correction capacitors?

- a. Shock through discharge energy.
- b. Shock from the supply cable.
- c. Burns from the outer casing.
- d. Fire from the capacitor electrolyte.

15 What would be a logical way to locate a short circuit fault on an isolated radial circuit having multiple socket-outlets?

- a. Start at the last socket and test at each socket working towards the Distribution Board.
- b. Start at the Distribution Board and test at each socket working towards the last socket.
- c. Start at a mid-point and determine the which section has the fault, then half that section again.
- d. Start by removing all sockets and twist all conductors together at each point before testing.

16 What test instrument is the most appropriate when locating an open circuit?

- a. Low resistance ohmmeter.
- b. Insulation resistance tester.
- c. Approved voltage indicator.
- d. Phase rotation indicator.

17 What test instrument would be used to confirm a suspected overload?

- a. Low resistance ohmmeter.
- b. Clamp on ammeter.
- c. Approved voltage indicator.
- d. RCD test instrument.

18 What procedure would ensure accurate results when using a GS 38 multimeter to measure voltage?

- a. Selecting an appropriate scale.
- b. Nulling the leads before use.
- c. Removing the probe fuses.
- d. Reversing the test leads.

19 What must be done to establish accurate readings when using a low resistance ohmmeter and a long wander lead?

- a. Test the meter on a known supply.
- b. Test the leads on a 9 V battery.
- c. Reverse the leads after the first test.
- d. Null or zero the test and wander leads.

20 What would be an expected test result when testing the continuity of a 30 m conductor which has a resistance of 7.41 mΩ/m?

- a. 0.074 Ω
- b. 0.222 Ω
- c. 0.741 Ω
- d. 2.220 Ω

21 What insulation resistance reading would indicate a healthy circuit without any concerns?

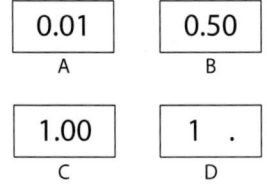

22 What has an effect on the downtime for a machine that has a broken drive belt?

- a. Cost of the new belt.
- b. Disposal of the old belt.
- c. Availability of the new belt.
- d. Service life of the old belt.

EXAM PRACTICE 8

EXAM PRACTICE 8: UNIT 014 UNDERSTAND FAULT DIAGNOSIS AND RECTIFICATION

23 What is the **main** consideration when repairs are being planned to a DNO distribution system supplying a hospital?

- a. Provision of standby generators.
- b. Provision of toilet facilities.
- c. Provision of water services.
- d. Provision of vehicle access.

24 What must be investigated if a new component fails?

- a. Manufacturer's instructions.
- b. Manufacturer's warranty conditions.
- c. Installer's qualifications.
- d. Installer's public liability insurance.

25 Who is responsible for the electrical installation in a college whilst electrical contractors are undertaking fault finding and diagnosis work?

- a. The Principle.
- b. The duty officer.
- c. The contractors.
- d. The duty-holder.

26 What test would confirm that a loose Neutral connection has been suitably repaired?

- a. Insulation resistance.
- b. Earth loop impedance.
- c. Conductor continuity.
- d. Phase sequencing.

27 What test would confirm that a single-phase socket-outlet has been correctly connected following repair?

- a. Polarity.
- b. Insulation.
- c. Phase sequence.
- d. Functional test.

Notes

28 What test result indicates that an RCD is not too sensitive following replacement?

- a. Trips within 300 ms at 1 x $I_{\Delta N}$.
- b. Trips within 40 ms at 1 x $I_{\Delta N}$.
- c. Trips within 500 ms at 0.5 x $I_{\Delta N}$.
- d. No trip within 2 s at 0.5 x $I_{\Delta N}$.

29 What does a phase sequence test establish before a replacement machine is switched on for the first time?

- a. The machine will rotate in the correct direction.
- b. The machine functions at the correct voltage.
- c. The machine has the correct overload setting.
- d. The machine has the correct type of starter.

30 What is the correct way to dispose of cardboard packaging?

- a. Burn in a bonfire.
- b. Place in a rubble skip.
- c. Take for recycling.
- d. Take to landfill.

Exam practice 8 questions with answers: Unit 014 understand fault diagnosis and rectification

1. What is the **most** likely reason an electrician received an electric shock whilst undertaking fault diagnosis?

   - a. No 30 mA RCD protection was present.
   - ● b. Safe isolation was not undertaken.
   - c. The fault tripped the circuit breaker.
   - d. The correct PPE was not present.

   Correct answer b
   If safe isolation of the installation or circuit where the fault is present is carried out, electric shock cannot realistically occur except for rare circumstances.

   One key mistake electricians can make is being so focussed on finding the fault that safe procedures become secondary. This should never happen.

2. What is the final method of protection when considering a risk assessment?

   - a. Safe and secure isolation.
   - ● b. Personal protective equipment.
   - c. Warning notices and signage.
   - d. Communication with personnel.

   Correct answer b
   When considering a risk assessment, personal protective equipment (PPE) must always be considered as the last line of defence and all other reasonable measures must be put in place first.

3. What are the minimum voltages where GS38 becomes applicable?

   - a. 50 V DC: 50 V AC
   - b. 50 V DC: 120 V AC
   - ● c. 120 V DC: 50 V AC
   - d. 120 V DC: 120 V AC

   Correct answer c
   GS 38 becomes applicable when testing voltages, or using test voltages, above 50 V AV or 120 V DC

Notes

4 What is the **most** useful document when establishing the correct operation of a specific item of equipment?

- ○ a. As fitted drawings and plans.
- ○ b. Electrical Installation Certificate.
- ○ c. Minor Electrical Works Certificate.
- ● d. Manufacturers' instruction manual.

Correct answer d
Manufacturers' instructions contain specific detail on the correct operation of equipment and should always be kept for referencing should the equipment become faulty.

5 What is the **main** advantage for communicating with users before undertaking fault diagnosis work to equipment?

- ● a. To ascertain the nature of the fault.
- ○ b. To establish the final cost of the repair work.
- ○ c. To ascertain the location of catering facilities.
- ○ d. To establish how to repair the equipment.

Correct answer a
Communicating with users of equipment can be very advantageous as a pattern can be ascertained of possible causes of a fault leading to efficient fault location and diagnosis.

6 What is the **most** likely reason for the continuous operation of an RCBO protecting a lighting circuit within a building being refurbished?

- ○ a. Overloading of the circuit.
- ● b. Damage to a concealed cable.
- ○ c. Corrosion in a light switch.
- ○ d. Loose earth connection in light.

Correct answer b
The most likely fault that has caused the continual operation of a protective device in this particular location is damage to a concealed cable as it is likely that the cable was damaged by a drill or fixing.

EXAM PRACTICE 8

EXAM PRACTICE 8 QUESTIONS WITH ANSWERS: UNIT 014 UNDERSTAND FAULT DIAGNOSIS AND RECTIFICATION

7 What is the **most** likely reason for an arcing sound in a light switch?

- a. Voltage drop.
- b. Open circuit.
- c. Short circuit.
- ● d. Loose connection.

Correct answer d
The most likely cause of an arcing sound is a loose connection making casual contact with a terminal creating an arc.

8 What causes a reduced voltage at a load?

- a. Short circuit lengths.
- b. Open circuit faults.
- ● c. Long circuit lengths.
- d. Short circuit faults.

Correct answer c
Voltage drop is commonly caused by long cable runs giving higher resistances. An open circuit will cause a loss of voltage as opposed to a reduction in voltage as detailed in the question.

9 What is the **most** likely cause when a circuit breaker operates, is re-set, then operates again after several minutes?

- a. Voltage drop.
- ● b. Overloading.
- c. Short circuit.
- d. Open circuit.

Correct answer b
When a circuit is overloaded, the thermal part of a circuit breaker will operate after several minutes. Once the circuit breaker cools, it can be re-set and will then trip again if the overloading continues.

10

Line-Line loop	0.4 Ω	Step 1 results
Neutral-Neutral loop	0.4 Ω	
CPC-CPC loop	0.4 Ω	
Readings at socket-outlet front	Open circuit	Step 2 results **Line-Neutral**
	0.2 Ω	Step 3 results **Line-CPC**

A continuity of ring-final circuit test is carried out.

What is the **most** likely fault at a socket-outlet given the readings in the table above?

- ⦿ a. Reversed polarity between Line and CPC.
- ○ b. A break in the line and the Neutral.
- ○ c. Reversed polarity between line and neutral.
- ○ d. A break in the neutral conductor.

Correct answer a

The results shown in the table would indicate a reverse polarity because:

- When the step 1 results were obtained, it proved there were no disconnected or breaks in the conductors removing A and D as possible answers.
- When Line to Neutral was tested at the socket-outlet front, an open circuit was obtained as the test was actually between CPC and neutral conductors.
- When Line to CPC was tested, the result was as expected even through the CPC was in the Line terminal and the Line was in the CPC terminal.
- If the fault was a Line and Neutral reversal, the expected reading would be established between Line and Neutral but an open circuit between Line and CPC.

EXAM PRACTICE 8

EXAM PRACTICE 8 QUESTIONS WITH ANSWERS: UNIT 014 UNDERSTAND FAULT DIAGNOSIS AND RECTIFICATION

Notes

11 What is the **most** likely reason for an RCD, controlling socket-outlets, to trip intermittently in a dwelling?

- a. A short in the wiring system.
- ● b. A faulty appliance such as a fridge.
- c. An open circuit in the wiring system.
- d. A loose bonding connection at the installation MET.

Correct answer b
Where an RCD trips from time to time and not constantly, it is likely an appliance is the problem. If the fault was in the wiring system, it is likely that the RCD would not re-set at all.

12 What location would create the greatest risk of electrocution due to a reduction in body resistance?

- a. Working on a wooden platform.
- ● b. Working in a metal ducting system.
- c. Working in a tiled bathroom.
- d. Working on an asphalt roof.

Correct answer b
When a person is within a metal duct the person is potentially in contact with multiple paths in parallel to earth. This creates a much lower body resistance meaning much lower voltages can be a risk of electrocution to that person.

13 What equipment is most vulnerable to damage by insulation resistance testing?

- a. Discharge lamps.
- b. Circuit wiring.
- ● c. Electronic controllers.
- d. Circuit breakers.

Correct answer c
The high voltages used when insulation testing can damage electronic equipment and this type of equipment should be disconnected before applying the test.

Notes

14 What is the **main** risk when working on isolated equipment having power factor correction capacitors?

- ● a. Shock through discharge energy.
- ○ b. Shock from the supply cable.
- ○ c. Burns from the outer casing.
- ○ d. Fire from the capacitor electrolyte.

Correct answer a
Equipment which has capacitors fitted must be discharged before working on them as the capacitors store a charge which can be released, causing electrocution, if contact was made with the equipment terminals.

15 What would be a logical way to locate a short circuit fault on an isolated radial circuit having multiple socket-outlets?

- ○ a. Start at the last socket and test at each socket working towards the Distribution Board.
- ○ b. Start at the Distribution Board and test at each socket working towards the last socket.
- ● c. Start at a mid-point and determine which section has the fault, then half that section again.
- ○ d. Start by removing all sockets and twist all conductors together at each point before testing.

Correct answer c
Logically, the circuit should be halved to determine the section where the fault is, then that section halved again to narrow down where the fault is. Although answers A and B would eventually locate the fault, they are not as efficient as C. Option D would not help locate the fault at all.

16 What test instrument is the most appropriate when locating an open circuit?

- ● a. Low resistance ohmmeter.
- ○ b. Insulation resistance tester.
- ○ c. Approved voltage indicator.
- ○ d. Phase rotation indicator.

Correct answer a
A low resistance ohmmeter is the safest and most appropriate instrument to test circuit continuity when locating a break in a circuit.

EXAM PRACTICE 8

EXAM PRACTICE 8 QUESTIONS WITH ANSWERS: UNIT 014 UNDERSTAND FAULT DIAGNOSIS AND RECTIFICATION

17 What test instrument would be used to confirm a suspected overload?

- a. Low resistance ohmmeter.
- ⦿ b. Clamp on ammeter.
- c. Approved voltage indicator.
- d. RCD test instrument.

Correct answer b
If overload in a circuit is suspected, the load current would be measured and compared to the circuit criteria such as the circuit rating.

18 What procedure would ensure accurate results when using a GS 38 multimeter to measure voltage?

- ⦿ a. Selecting an appropriate scale.
- b. Nulling the leads before use.
- c. Removing the probe fuses.
- d. Reversing the test leads.

Correct answer a
By selecting a scale closest to the anticipated voltage, a more accurate value would be displayed by the meter.

19 What must be done to establish accurate readings when using a low resistance ohmmeter and a long wander lead?

- a. Test the meter on a known supply.
- b. Test the leads on a 9 V battery.
- c. Reverse the leads after the first test.
- ⦿ d. Null or zero the test and wander leads.

Correct answer d
When using any test leads and a wander lead with a low resistance ohmmeter, the resistance of all leads must be nulled or zeroed to remove those resistances from any reading.

Notes

Notes

20 What would be an expected test result when testing the continuity of a 30 m conductor which has a resistance of 7.41 mΩ/m?

- ○ a. 0.074 Ω
- ● b. 0.222 Ω
- ○ c. 0.741 Ω
- ○ d. 2.220 Ω

Correct answer b
The expected result would be calculated by

$$\frac{\text{mΩ/m} \times \text{length}}{1000} = R$$

so

$$\frac{7.41 \times 30}{1000} = 0.222 \, \Omega$$

21 What insulation resistance reading would indicate a healthy circuit without any concerns?

0.01	0.50
A	B

1.00	1 .
C	D

Correct answer d
A healthy circuit with no concerns of damage or deterioration would be indicated by an insulation resistance value over and above the measuring scale of the instrument. This would be indicated by the display shown in D.

Although option C is an acceptable value of insulation resistance, concerns should be raised as to why it is as low as 1 MΩ.

EXAM PRACTICE 8

EXAM PRACTICE 8 QUESTIONS WITH ANSWERS: UNIT 014 UNDERSTAND FAULT DIAGNOSIS AND RECTIFICATION

22 What has an effect on the downtime for a machine that has a broken drive belt?

- a. Cost of the new belt.
- b. Disposal of the old belt.
- ● c. Availability of the new belt.
- d. Service life of the old belt.

Correct answer c
The availability of the required new belt directly affects the time the machine will be out of action.

23 What is the **main** consideration when repairs are being planned to a DNO distribution system supplying a hospital?

- ● a. Provision of standby generators.
- b. Provision of toilet facilities.
- c. Provision of water services.
- d. Provision of vehicle access.

Correct answer a
If the supply to an important building, such as a hospital, is planned to be off, provision of standby generators must be considered.

24 What must be investigated if a new component fails?

- a. Manufacturer's instructions.
- ● b. Manufacturer's warranty conditions.
- c. Installer's qualifications
- d. Installer's public liability insurance.

Correct answer b
If a new component fails the warranty agreement and conditions should be checked and the part exchanged.

Notes

Notes

25 Who is responsible for the electrical installation in a college whilst electrical contractors are undertaking fault finding and diagnosis work?

- a. The Principle.
- b. The duty officer.
- c. The contractors.
- ● d. The duty-holder.

Correct answer d
Whilst electrical work is undertaken in a building, such as a college, one person, usually the person who is supervisor for the electrical contractors, becomes the Duty Holder and is responsible for the health and safety of all persons within the location of the work.

26 What test would confirm that a loose Neutral connection has been suitably repaired?

- a. Insulation resistance.
- b. Earth loop impedance.
- ● c. Conductor continuity.
- d. Phase sequencing.

Correct answer c
If the repair to a loose connection is correct, continuity would be restored.

27 What test would confirm that a single-phase socket-outlet has been correctly connected following repair?

- ● a. Polarity.
- b. Insulation.
- c. Phase sequence.
- d. Functional test.

Correct answer a
Polarity testing ensures that the connection of equipment is correct and should be verified before the socket is used.

EXAM PRACTICE 8

EXAM PRACTICE 8 QUESTIONS WITH ANSWERS: UNIT 014 UNDERSTAND FAULT DIAGNOSIS AND RECTIFICATION

28 What test result indicates that an RCD is not too sensitive following replacement?

- a. Trips within 300 ms at $1 \times I_{\Delta N}$.
- b. Trips within 40 ms at $1 \times I_{\Delta N}$.
- c. Trips within 500 ms at $0.5 \times I_{\Delta N}$.
- ● d. No trip within 2 s at $0.5 \times I_{\Delta N}$.

Correct answer d
When an RCD is tested using a current half the residual setting of the RCD. The device should not trip within 2 seconds. If it does, it is an indication that the RCD may be too sensitive.

29 What does a phase sequence test establish before a replacement machine is switched on for the first time?

- ● a. The machine will rotate in the correct direction.
- b. The machine functions at the correct voltage.
- c. The machine has the correct overload setting.
- d. The machine has the correct type of starter.

Correct answer a
Phase sequence testing ensures that three-phase machines rotate in the correct direction.

30 What is the correct way to dispose of cardboard packaging?

- a. Burn in a bonfire.
- b. Place in a rubble skip.
- ● c. Take for recycling.
- d. Take to landfill.

Correct answer c
Cardboard packaging should always be recycled even if other methods of disposal many be more convenient.

Notes

Exam practice 8 answer key: Unit 014 understand fault diagnosis and rectification

1 b	7 d	13 c	19 d	25 d
2 b	8 c	14 a	20 b	26 c
3 c	9 b	15 c	21 d	27 a
4 d	10 a	16 a	22 c	28 d
5 a	11 b	17 b	23 a	29 a
6 b	12 b	18 a	24 b	30 c

More information

Further reading 186

Online resources 187

Further qualifications 188

Further reading

We recommend that apprentices ensure they are familiar with the latest editions of:

BS 7671 Requirements for Electrical Installations, IET Wiring Regulations. This document is amended and re-published every three years.

Other useful publications aimed at electrical apprentices are:

Student's Guide to Calculations, published by the IET
Student's Guide to the IET Wiring Regulations, published by the IET

Other useful publications, which are updated in line with BS 7671 include:

On-Site Guide: BS 7671, published by the IET
Electrician's Guide to the Building Regulations, published by the IET

IET Guidance Notes also offer further guidance, expanding on particular requirements of the Wiring Regulations. They are published by the IET.

Guidance Note 1: Selection and Erection
Guidance Note 2: Isolation and Switching
Guidance Note 3: Inspection and Testing
Guidance Note 4 Protection against Fire
Guidance Note 5: Protection against Electric Shock
Guidance Note 6: Protection against Overcurrent
Guidance Note 7: Special Locations
Guidance Note 8: Earthing and Bonding

MORE INFORMATION
ONLINE RESOURCES

Online resources

Notes

City & Guilds www.cityandguilds.com

The City & Guilds website can give you more information about studying for further professional and vocational qualifications to advance your personal or career development as well as locations of centres that provide the courses.

SmartScreen www.smartscreen.co.uk

City & Guilds' dedicated online support portal SmartScreen provides learner and tutor support for over 100 City & Guilds qualifications. It helps engage learners in the excitement of learning and enables tutors to free up more time to do what they love most – teach!

Institution of Engineering and Technology (IET) www.iet.org

The Institution of Engineering and Technology is the largest professional engineering society in Europe and the second largest of its kind in the world. The Institution produces the Wiring Regulations in conjunction with the British Standards Institution and a range of supporting material which can be accessed in hard copy or electronically.

* British Standards Institution www.bsi-global.com

* Electrotechnical Certification Scheme (ECS) www.ecscard.org.uk

    ECS administers the card scheme for electrical professionals.

* ELECSA Ltd www.elecsa.co.uk

* Joint Industry Board www.jib.org.uk

* Electrical Safety First www.electricalsafetyfirst.org.uk

* National Inspection Council for Electrical Installation Contracting (NICEIC) www.niceic.com

Further qualifications

2391 City & Guilds Level 3 Award in Inspection and Testing

Once you have completed the 5357 qualification and gained some site experience, you are eligible to enrol on the 2391-51 qualification on Periodic inspection and Testing. Within your 5357 qualification, you received sufficient training and you examination was to a standard meaning you do not need to sit the 2391-50 examination for Initial Verification.

2382 City & Guilds Level 3 Certificate in the Requirements for Electrical Installations BS 7671

This qualification is aimed at practicing electricians and is intended to ensure they are conversant with the content of the current IET Wiring Regulations. If you complete 5357, you will have already gained recognition for this examination but always remember to keep up to date which you can do by taking this examination when BS 7671 is amended.

2377 City & Guilds Level 3 Certificate in the Code of Practice for the In-service Inspection and Testing of Electrical Equipment

This course, commonly known as PAT/Portable appliance testing, is for operatives undertaking and recording or managing electrical equipment which is not part of a fixed electrical installation.

2392 City & Guild Level 2 Certificate in Fundamental Inspection, Testing and Initial Verification

This qualification provides inexperienced electricians an introduction to the initial verification and certification process. It is intended to prepare candidates to go on to the Level 3 Award in Inspection and Testing (2391).

2393 City & Guilds Level 3 Certificate in the Building Regulations for Electrical Installations

This qualification is intended for practicing electricians working in domestic dwellings who need to be conversant with relevant parts of the Building Regulations.

MORE INFORMATION
FURTHER QUALIFICATIONS

2396 City Guilds Level 4 Award in the Design, Erection and Verification of Electrical Installations

This qualification is intended for experienced persons working in the Electrical Installation industry and provide them with the skills and knowledge of electrical installation design in order to comply with BS 7671.

2919 City & Guilds Level 3 Award in Electric Vehicle Charging Installations

This qualification is intended for experienced electricians wishing to undertake electric vehicle charging installation work in line with the IET Code of Practice in Electric Vehicle Charging Installations.

2356 City & Guilds Level 3 Qualification in Electrotechnical Services and Systems

This qualification is intended for those working as an electrician but have not got the necessary formal NVQ to prove competency to the National Occupational Standards required for registration of the Electrical Certification Scheme (ECS).

For more information on the suite of Electrical Qualifications in the Building Services Portfolio visit http://www.cityandguilds.com/qualifications-and-apprenticeships/building-services-industry

Notes

Notes

Electrical resources for learners

The IET provides a wealth of materials to support learners studying for electrotechnical qualifications.

The Student's Hub on the IET website offers you:

Videos – giving an insight into different areas of BS 7671, as well as looking at some of the tools most used in day-to-day electrical work

The Job Profile Bank – providing an overview of some of the exciting possibilities available for those entering the electrical industry

Practical study guides – such as the Student's Guide to the IET Wiring Regulations, a handy companion to those studying for BS 7671 qualifications

www.theiet.org/students-hub

You can also find all of the Exam Preparation series of books, as well as BS 7671 (IET Wiring Regulations) and all of the expert IET guidance at **www.theiet.org/wiringbooks**

The Institution of Engineering and Technology is registered as a Charity in England & Wales (no 211014) and Scotland (no SC038698). The IET, Michael Faraday House, Six Hills Way, Stevenage, SG1 2AY, UK.

Notes